Analysis of Oil and Gas Production Performance

Analysis of Oil and Gas Production Performance

Steven W. Poston, Texas A&M University (Retired)

Marcelo Laprea-Bigott, Texas A&M University

Bobby D. Poe Jr., Schlumberger (Retired)

Society of Petroleum Engineers

© Copyright 2019 Society of Petroleum Engineers

All rights reserved. No portion of this book may be reproduced in any form or by any means, including electronic storage and retrieval systems, except by explicit, prior written permission of the publisher except for brief passages excerpted for review and critical purposes.

Printed in the United States of America.

Disclaimer

This book was prepared by members of the Society of Petroleum Engineers and their well-qualified colleagues from material published in the recognized technical literature and from their own individual experience and expertise. While the material presented is believed to be based on sound technical knowledge, neither the Society of Petroleum Engineers nor any of the authors or editors herein provide a warranty either expressed or implied in its application. Correspondingly, the discussion of materials, methods, or techniques that may be covered by patents implies no freedom to use such materials, methods, or techniques without permission through appropriate licensing. Nothing described within this book should be construed to lessen the need to apply sound engineering judgement nor to carefully apply accepted engineering practices in the design, implementation, or application of the techniques described herein.

ISBN 978-1-61399-665-2

First Printing 2019

Society of Petroleum Engineers
222 Palisades Creek Drive
Richardson, TX 75080-2040 USA

https://www.spe.org/store
books@spe.org
1.972.952.9393

Dedication

This book is dedicated to the past, present, and future petroleum-engineering students passing through Texas A&M University.

Additionally, R. L. Whiting and D. von Gonten, former heads of the Department of Petroleum Engineering at Texas A&M University, should always be remembered for their foresight and efforts to expand and extend the scope of the department to the parents, state, and industry. R. Berg and R. Morse were of a great guidance and help while we were at the department.

Preface

Oil and gas companies are continually seeking and applying new technologies, processes, and methods to reduce their cost of finding and producing hydrocarbons while remaining competitive in the current and changing global economy. Improving the efficiency of business processes and maximizing the productivity of the workforce will help to reduce the associated costs and should ultimately increase profitability.

Although technology has helped companies to better evaluate the prospects, lack of trained geoscientists and engineers and the absence of proper vehicles for training and technology transfer may jeopardize oil- and gasfield-development efforts. An efficient and effective way to help develop core competencies for different jobs is to design tools and training actions to address these needs. In this book, workflows have been developed that apply key technology independently by analyzing the processes and solving example problems, thereby addressing the importance of integration of subsurface disciplines related to oil and gas exploration and production.

Traditional Arp's models exist that are based on graphical extrapolation of production data, and they have been regarded in our industry as one of the preferred and commonly-used tools for estimating future performance in oil and gas wells. However, the practical aspects of analyzing production performance have changed as a result of the increased exploitation efforts in unconventional reservoirs. The complexities of these types of reservoirs were not adequately covered in the initial work *Analysis of Production Decline Curves*, published by the Society of Petroleum Engineers in October 2008. In the current book, the scope has been broadened, and we provide many more field examples, including problems that cover the specific subjects of developing well-evaluation procedures and best practices for new areas of shale and tight formation reservoirs.

Advances in horizontal well drilling and multistage hydraulic fracturing have allowed industry to develop unconventional nano-Darcy permeability reservoirs (shale oil and shale gas). These highly-heterogeneous multiphase systems do not lend themselves to typical analytical solutions to predict future performance. Boundary conditions applied to such systems are based on ideal geometrical configurations and idealized flow theory. This approach implies important and sometimes faulty assumptions concerning geological heterogeneity and multiphase flow in the physical system. Aspects of production forecasting in unconventional resources are now covered in this book. The sections discussing type curve and two phase flow have been expanded and revised completely, and an additional section on types of equations replicating different flow conditions encountered in the oil field is presented. The most useful plotting and interpretive methods have been added, and a method for estimating ultimate recovery is included.

This book is intended for engineers, geologists, and anyone working in the oil and gas industry with an interest in production forecasting of conventional and unconventional resources for evaluation and development. The majority of the book is concerned with commonly observed oilfield practice and practical solutions to the problems encountered therein. Each chapter begins with a workflow diagram that, in essence, provides the reader with the learning objectives of the chapter. A primary focus of the book is to instill each reader with the competency to solve typical operational problems with minimal exposure to the complexity of the underlying mathematics and equations. The basics and utility of each equation are discussed; however, the focus is on the practical application of the underlying technology to real-life problems. There are numerous illustrations and solutions to typical field problems included for the reader.

About the Authors

Steven W. Poston is a retired professor emeritus from Texas A&M University, with more than 18 years of teaching experience at the graduate and undergraduate levels in applied reservoir engineering analysis and subsurface description of petroleum reservoirs both in the United States and internationally. His industry experience includes more than 14 years in a variety of subsurface engineering, geological, and managerial roles for Gulf Oil Exploration and Production Company in Nigeria, Pennsylvania, Louisiana, and Texas. Poston has coauthored numerous industry technical papers, has served on a number of SPE committees, and was coauthor of the SPE books *Overpressured Gas Reservoirs* and *Analysis of Production Decline Curves*.

Marcelo Laprea-Bigott is professor of engineering practice at the Harold Vance Department of Petroleum Engineering at Texas A&M University and has more than 45 years of consulting experience and an extensive background in developing and delivering technical training courses in the field of reservoir engineering for a variety of national oil companies and international clients. He joined Texas A&M University in 2006 after 20 years with Schlumberger Data and Consulting Services, having been an advisor in their Network of Excellence in Training (NExT). Laprea-Bigott's other private sector industry experience includes more than 13 years as founder, principal owner, and president of Simupet, C.A., a petroleum engineering consulting firm; Consorcio Lamar C.A., an integrated oilfield tubulars management company; and president of S. A. Holditch and Associates–Venezuela. Additionally, he was a professor at Universidad de Oriente–Venezuela, a visiting adjunct professor at Texas A&M University, a visiting adjunct professor at the University of Tulsa, and the former assistant director of the Energy Institute at Texas A&M University. Laprea-Bigott holds a PhD degree from Texas A&M University.

Bobby D. Poe Jr. is a production and reservoir engineer with more than 30 years of industrial experience in all levels of well stimulation, well performance, and reservoir-engineering analyses. His most recent engineering position is as the production interpretation advisor at Schlumberger in Houston. Poe holds a PhD degree in petroleum engineering from Texas A&M University. He has served on a number of SPE Reprint Series and editorial committees, has authored or coauthored numerous technical articles related to production analyses, and coauthored the SPE book *Analysis of Production Decline Curves*. Poe holds more than a dozen US and international patents related to production-engineering analyses and interpretation techniques.

Table of Contents

Preface vii
About the Authors ix
Chapter 1 – Introduction to Decline Curves 1
    Introduction to Decline Curves 1
    Physical Considerations 2
    Arps Equations 4
    Transient Boundary-Dominated Conditions 12
Chapter 2 – Effects of Field Conditions 19
    Field Examples 20
    Production Segments 21
    Correcting for Well Downtime 23
    Informational Plots 24
    Multiplot Analysis 26
Chapter 3 – Smoothing Variable Production 31
    Production Histories 31
    Pseudo Production Time 31
    Normalizing Production 34
    The Quadratic Model 35
Chapter 4 – Well and Reservoir Models 47
    Geological Considerations 48
    Unfractured Vertical Wellbore Models 48
    Horizontal-Unfractured-Well Case 51
    Horizontal Well Example 52
    Fractured Vertical Wellbore Case 55
Chapter 5 – Fractured Horizontal Wells 67
    Geological Setting 67
    Depletion Model 71
    Shale Well Examples 74
    Regional Analysis Example 78
Chapter 6 – Type Curves 85
    Introduction 85
    The Problem 86
    Fetkovich Method 87
    The Blasingame et al. Type Curve Method 97
    Poe and Poston Type Curves 102
Chapter 7 – Two Phase Flow 115
    Geological Influences 116
    The Waterflood Event 117
    Performance History—Provost Field 122
    Estimating Reserves and Predicting Performance 123
    Analysis Procedure 124
    Analysis Procedure 126
    Results of Analyses 126
    Relative Permeability 127
    Multiple Performance Plots 135
    Well Diagnostics Plots 142
Nomenclature 155
References 159
Index 163

Chapter 1

Introduction to Decline Curves

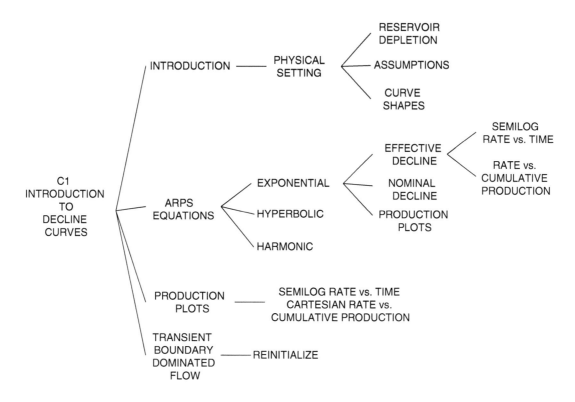

The first chapter sets the foundation for the ensuing work, which delves deeply into different aspects of production decline analysis. The study of production performance is often denigrated because of the often-uncertain data quality. However, it is the one set of data most often available for estimating well character.

Every well or field does not lend itself to decline curve analysis. The reader is initially introduced to some of these general uncertainties and assumptions. One should be aware of these fundamentals before analyzing and predicting performance no matter how sophisticated the approach is.

Mathematically fitting an equation of a line to a production decline curve has been attempted by various authors in the past. However, Arps (1945) was the first to present a unified approach for analyzing a performance curve. Because of the complexity of the analysis process, estimating future performance with the Arps hyperbolic equation was not widely pursued until the advent of personal computers.

We will review these fundamentals in this chapter.

Introduction to Decline Curves

Production decline curve analysis is a classical reservoir engineering technique, applicable to both oil and gas wells. Production decline analysis is a traditional means of identifying well production problems and predicting well performance and well life on the basis of real production data. The decline curve analysis predictions are valid only if factors that influenced the performance trends of wells or fields in the past would continue to govern their performance in the same manner.

Oil and gas production rates generally decline as a function of time. Chief factors in the decline are discussed next.

Fitting a line through declining production values and assuming this same line trends forward in a similar manner forms the basis for analyzing decline curves. However, similarity of current and future performance is not necessarily a function of the equation of a line. In fact, the character production curve is derived from of the rock fabric, fluid type, completion characteristics and producing rate.

It has some important and generic applications:

- Can be conducted on well, reservoir, and field level
- Can be used to determine the reserves for a well, lease, or field
- Independent method of reserves estimation, the result of which on conventional reservoirs can be compared with volumetric or material-balance estimates
- Can be performed to estimate a base line to evaluate the success of future production enhancement (i.e. Future infill drilling, fluid injection, fracturing, acidizing) operations
- Can be used for the evaluation of new investments; audit of previous expenditures; sizing of equipment and facilities such as pipelines, plants, and treating facilities

Arps (1945) introduced the first systematic approach for the analysis of decline curves by empirical methods. Fetkovich (1980) introduced type curves and methodology to analyze transient and boundary-dominated flow periods, and Blasingame et al. (1991) and Agarwal et al. (1998) published work for using type curves and derivative curves accounting for flowing pressure variations.

Later work concentrated on the application of production decline analysis to fractured unconventional oil and gas systems. The classical Arp's approach uses empirical models with little fundamental justification and uses only production data, (no special reservoir parameters are required) and gives

- Forecast of future production rates
- Reserves estimation

Modern techniques involve a theoretical approach and account for pressure variations and reservoir parameters. Advanced decline curve analysis gives

- Estimation of k and S
- Distinction between transient and boundary flow
- Forecast of future production rates
- Reserves and original-oil-in-place (OOIP) and original-gas-in-place (OGIP) estimations

Chief factors for the oil and gas production rates decline as a function of time are

- Reservoir pressure provides energy to drive fluids from the reservoir (p_{res}) to the perforations (p_{wf}), and then to the surface (p_{tf}). Continued depletion of oil or gas fluids causes loss of reservoir pressure, which in turn affects production rate.
- Changing relative volumes of produced fluids. An unwanted fluid, such as water or gas in the case of an oil well or water in the case of a gas well, enters the flow stream. Decreased production of the primary product is the result of the onset of two-phase production and increased hydrostatic head.

Other frequent possible factors are

- Increase in near-wellbore damage (Skin>0)
- Production problems (e.g., sand production, scale, asphaltenes)

Fitting a line through declining production values and assuming this same line trends forward in a similar manner forms the basis for analyzing decline curves. However, similarity of current and future performance is not necessarily a function of the equation of a line. In fact, the character of the production curve is derived from of the rock fabric, fluid type, completion characteristics, and producing rate.

Physical Considerations

Production rates initially are dependent on growth of the expanding drainage system. Depletion is a function of an apparent increasing drainage volume (infinite-acting flow behavior also known as transient flow). On the other hand, encountering a reservoir boundary implies production is controlled by the drainage volume (boundary dominated flow). Including effects of infinite-acting flow implies an increasing reserves estimate. This fact presents a particular problem when studying very-low-permeability reservoirs.

Rocks are seldom distributed in a homogeneous manner but are often layered during the sedimentary process. Each layer is composed of rocks of different properties and furnishes different depletion rates to the flow stream. The expansion rate of a disturbance migrating outward from the wells is based on the diffusion constant $\left(\eta = \dfrac{k}{\phi \mu c_t}\right)$, where k = permeability, ϕ = porosity, μ = viscosity, and c_t = total compressibility.

One can see that a thousandfold difference in permeability could materially affect production rates from a low-permeability or layered sand. Including production derived from natural or hydraulic fractures would add further complexity of analysis because of their dual permeability.

Reservoir Depletion. **Fig. 1.1** illustrates expansion of reservoir drainage limits from inception when the well is placed on production until an outer boundary (r_e) is encountered. The well is operating under constant flowing bottomhole-pressure (p_{wf}) conditions.

Boundary-Dominated Flow. The equation for calculating time required for a reservoir to transition from infinite-acting to boundary dominated flow conditions is:

$$t_{pss} \approx \frac{40 \phi \mu c_t r_e^2}{k} \quad \dotfill \quad (1.1)$$

where t_{pss} is in days.

Table 1.1 shows time required for a pressure disturbance to travel from the wellbore to the outer boundary for three reservoir cases. These calculations show boundary dominated flow can be initiated in a matter of a few days

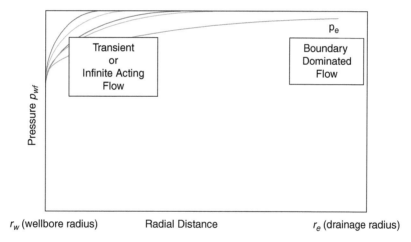

Fig. 1.1—Differentiating between constantly expanding (transient) and constant-volume (boundary dominated) conditions. An expanding drainage radius indicates an increasing reservoir volume. The pressure drop between (r_w) and (r_e) must begin to decline because a closed outer boundary has been encountered at boundary dominated conditions.

	Oil Reservoir	High-Pressure Gas Reservoir	Low-Pressure Gas Reservoir
Drainage area (acres)	160	640	640
Drainage area (sq ft)	1,490	2,980	2,980
Viscosity (cp)	0.6	0.022	0.018
Porosity (%)	12	12	12
Compressibility, 1/psi	20×10^{-6}	40×10^{-6}	170×10^{-6}
Permeability, md	50	10	100
t_{pss}, days	2.6	3.8	1.3

Table 1.1—Onset comparison of boundary conditions.

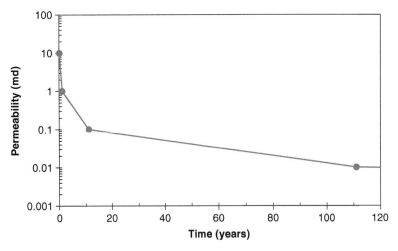

Fig. 1.2—Transient conditions can last for an extraordinarily long time for the very-low-permeability gas case.

for moderate-permeability and moderate-compressibility reservoirs. Also, the lower the compressibility, (the less gassy), the sooner boundary effects are encountered.

The more permeable layer in a noncommunicating, multizone completion becomes affected by the outer boundary within a shorter length of time than the less-permeable layers.

In conclusion, we can say that changing boundary conditions in variable-permeability reservoirs can cause the b-exponent to remain at high values and appear transient in nature, though it eventually trends toward zero.

Low-Pressure Gas Example. **Fig. 1.2** represents time in years required for a low-pressure (1,000 psia) dry gas reservoir to reach boundary-dominated-flow conditions as a function of average permeability. It is apparent that well life, although probably producing at a low rate for the very-low-permeability case, can last for years.

Assumptions. The following assumptions anticipate that a production history follows an unaltered and smooth decline. However, operational variations often divide a production history into segments, each reflecting different constant bottomhole pressure and production rate. These relations may be caused by

- The assumption that well flow is not mechanically restricted by chokes or tubing size. In actuality, production records often do not record choke changes. Dramatic flow rate changes signify something.
- Reservoir depletion conditions that remain relatively constant. Operational changes such as completing or abandoning wells might alter well drainage area which in turn can change performance characteristics. One should be careful about evaluating a production curve extending over a long period if history is not smoothly declining.
- Sufficient production performance data are available, and a declining trend has been established under boundary-dominated flow conditions; i.e., the well is draining a constant drainage area (pseudosteady flow if bottomhole pressure is constant).
- The well is produced at or near capacity. The productivity index of the well does not change. Factors that influenced the performance trends of wells or fields in the past will continue to govern their performance in the same manner.
- Absence of water influx or gas-cap expansion. Addition of an extra energy source must be considered when predicting future behavior.

Shapes of Production Decline Curves. Fig. 1.3 shows semilog rated time decline curves for four different wells located in the same field, but of different producing abilities. Line curvature defines future performance.

Fig. 1.4 shows how fitting a series of straight lines between two data points can provide a basis for predicting future production. The slopes of the declining straight lines constantly decrease.

Arps Equations

Arps (1945) modeled the various average shape of a line concepts to form a unified approach. The location of the fitted curve is defined in space by three values to form the equation of a line. These values are

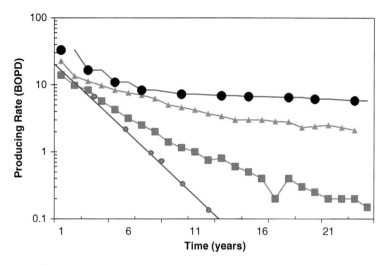

Fig. 1.3—Definitive shapes defined by the Arps equation. It can be observed that the top two curves never decline to a zero production rate, indicating transient flow conditions possible because of commingled layers (no crossflow) and Arps should be applied with caution. Decline curves are normally presented on a semilog rate vs. time plot.

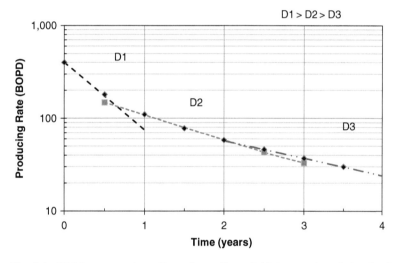

Fig. 1.4—Divide a curved semilog rate vs. time plot into a series of chords of decreasing slope. The declining straight-line slopes form a smoothly declining curve in the case for Fig. 1.2.

- Some initial producing rate (q_i)
- An initial decline rate (d_i) (might or might not coincide with the field data)
- The degree of curvature of the declining line (a function of the (b-exponent) term)

Arps (1945) defined the loss ratio as $-a = \dfrac{q}{\dfrac{dq}{dt}}$ or $a = -\dfrac{q}{\dfrac{dq}{dt}}$ . . . . (1.2)

The reciprocal of the loss ratio is defined as decline rate (D). where $a = 1/D$.

Therefore, $-\dfrac{1}{D} = \dfrac{q}{\dfrac{dq}{dt}}$ . . . . (1.3)

The loss ratio derivative, (the tangent slope of the line) is (b), where

$$b = \frac{d(1/D)}{dt} = \frac{d}{dt}\left(\frac{q}{dq/dt}\right). \quad\quad\quad\quad\quad\quad\quad\quad\quad\quad\quad\quad\quad\quad\quad\quad\quad (1.4)$$

Arps (1945) further defined that an exponential curve occurs when a series of rate/time estimates exhibit a constant (b) value and a hyperbolic curve when the derivative of the loss ratio remains constant.

Integrating over time results in a relationship between time, changing decline rates, and the b-exponent term:

$$D = \frac{D_i}{1+bD_i t}. \quad\quad (1.5)$$

Substituting the Arps definition given in Eq. 1.3 into previous equation 1.5 results in

$$-\frac{d(\ln q)}{dt} = \frac{D_i}{1+bD_i t} \quad\quad (1.6)$$

Substituting the Arps definition and integrating from $(0 \rightarrow t)$ develops an exponential rate vs. time expression

$$q_2 = q_1 \exp(-Dt) \quad\quad (1.7)$$

Please note $b = 0$. Integrating over $(0 \rightarrow t)$ develops a hyperbolic rate-time expression:

$$q_2 = \frac{q_1}{(1+btD_i)^{1/b}}. \quad\quad (1.8)$$

Eq. 1.8 reduces to Eq. 1.9 for the harmonic case,

$$q_2 = \frac{q_1}{(1+tD_i)} \quad\quad (1.9)$$

Please note $b = 1$.

Exponential Decline. The following develops two relationships for the exponential decline.

Constant Percentage Exponential Decline. Apply a stepwise definition for an exponential decline. The effective or constant rate decline expresses incremental rate loss as a stepwise function. Define the first rate as (q_1) and a subsequent rate as (q_2). The rate differences usually span 1 year. Be wary of a decline rate expressed in a lesser time span.

$$d = \frac{q_1 - q_2}{q_1} = -\frac{\frac{\Delta q}{\Delta t}}{q_1} = 1/\text{time}. \quad\quad (1.10)$$

Rearrange to $t = \dfrac{\ln \dfrac{q_2}{q_1}}{-\ln(1-d)}. \quad\quad (1.11)$

Rearrange to develop a rate equation:

$$q_2 = q_1(1-d)^t. \quad\quad (1.12)$$

Integrate from $(t_i \rightarrow t_2)$ and obtain cumulative production:

$$Q_p = \frac{q_1 - q_2}{-\ln(1-d)}. \quad\quad (1.13)$$

Convention assumes the decline rate is expressed in terms of %/yr.

Decline rates expressed in monthly units might be a subterfuge to force a well exhibiting a dramatic production falloff to appear in a better light.

Including the b-exponent term presents a major problem when adjusting the time-unit span. Monthly and daily decline rate equations are:

Convert from rate/year to rate/month:

$$d_y = 1 - (1-d_m)^{12}. \quad\quad (1.14a)$$

Convert from rate/year to rate/day:

$$d_y = 1 - (1-d_d)^{365}. \quad\quad (1.14b)$$

Example. When 220 BOPD is 12 months and 63 BOPD is 24 months are interpreted from a performance curve, calculate constant percentage decline rate.

$$d = \frac{q_1 - q_2}{q_1} = \frac{(220 - 63)(100)}{220} = 71.4\% / \text{yr}$$

Convert the decline rate time units from %/yr to %/month:

$$(1 - 0.714) = (1 - d_m)^{12}, \text{ or } d_m = (0.099)(100) = 9.9\% / \text{month}$$

Arps Nominal Decline. Arps nominal or continuous rate decline is considered here.

Arps (1945) and Brons* (personal communication) expressed the rate of change in the flow rate as a function of decline rate (D). Rearrange the exponential rate equation (Eq. 1.11) to solve for the decline rate:

$$D = \frac{\ln(q_1/q_2)}{t} \quad \quad (1.15)$$

Rearrange to provide a time interval relationship:

$$t = \frac{\ln(q_1/q_2)}{D} \quad \quad (1.16)$$

Substituting rate expression and integrating over integral limits 0 to t result in a cumulative production expression:

$$Q_p = \frac{q_1 - q_2}{D} \quad \quad (1.17)$$

Use estimated ultimate recovery (EUR) to estimate the theoretical maximum reserves. $EUR = Q_{max}$; assume $q_{last} = 0$ in the limit as t goes to infinity, for exponential decline. Please note this is not the economic limit (EL).

$$Q_{max} = EUR = \frac{q_1}{D} \quad \quad (1.18)$$

Comparing Constant and Continuous Declines. A rewritten form of the effective decline definition is

$$\frac{q_2}{q_1} = 1 - d \quad \quad (1.19)$$

Rewrite nominal decline definition as $\frac{q_2}{q_1} = \exp(-Dt)$. \quad (1.20)

Combining results in

$$d = 1 - \exp(-Dt) \quad \quad (1.21a)$$

Or, conversely:

$$D = -\ln(1 - d) \quad \quad (1.21b)$$

This development shows that the exponential decline definitions are different but will produce similar answers if the proper equations are applied.

The solid line in **Fig. 1.5** compares relative decline rate values for the constant percentage and continuous decline definitions. A 45° slope existing up to a 25% decline reflects a similarity between the two different methods. However, the continuous decline rate increases quite dramatically when compared to the constant percentage decline rate values after this point.

In conclusion, we see the exponential curve may be defined in the context of an effective or a nominal decline. The equations are different, but the results of the calculations are the same. Either can be applied to study exponential decline curves if the proper equation sets are applied.

---

*Brons, F. Personal Communication, 1966.

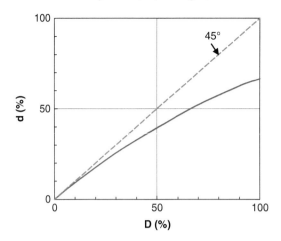

Fig. 1.5—Comparing effective and continuous decline rates. Note the similarity of the values up to approximately $D = 25\%$ value.

	Effective (Constant Percentage)	Continuous (Nominal)
Decline rate	$d = \dfrac{q_1 - q_2}{q_1}$	$D = \dfrac{\ln\left(\dfrac{q_1}{q_2}\right)}{t}$
Producing rate	$q_2 = q_1(1-d)^t$	$q_2 = q_1 \exp(-Dt)$
Elapsed time	$t = \dfrac{\ln\left(\dfrac{q_2}{q_1}\right)}{-\ln(1-d)}$	$t = \dfrac{\ln\left(\dfrac{q_1}{q_2}\right)}{D}$
Cumulative recovery	$Q_p = \dfrac{q_1 - q_2}{-\ln(1-d)}$	$Q_p = \dfrac{q_1 - q_2}{D}$
EUR	$Q_p = \dfrac{q_1}{-\ln(1-d)}$	$Q_p = \dfrac{q_1}{D}$

Table 1.2—Comparison of Effective and Continuous Decline Equations.

Rate vs. Time Plot. Express the exponential rate equation in logarithmic terms and arrange in the form of the equation of a straight line. See **Fig. 1.6.**

$$\ln q_2 = -Dt + \ln q_i \quad \text{..} \quad (1.22)$$

Components	
Plotting Variables	Outcome Variables
"y" axes: ($\ln q_2$)	"y" intercept: ($\ln q_i$)
"x" axis: (t)	the slope of the line is: ($-D$)

Table 1.3—Components of Rate vs. Time Plot Exponential decline.

Rate vs. Cumulative Production Plot. Rearrange the cumulative production equation to the equation of a straight line:

$$q_2 = -Q_p D + q_i \quad \text{..} \quad (1.23)$$

Components	
Plotting Variables	Outcome Variables
"y" axes: (q_2)	"y" intercept: (q_i)
"x" axis: (Q_p)	slope of the line is: ($-D$)

Table 1.4—Components of Rate vs. Cumulative Production Plot Exponential decline.

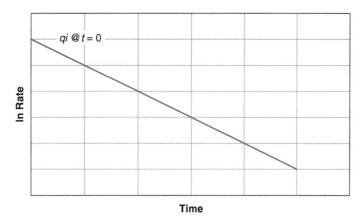

Fig. 1.6—The well-known exponential logarithmic rate vs. time plot which is the usual initial plot for all decline curve analysis. Predict future performance by extrapolating along the straight line. Note value of *qi* at *t* = 0.

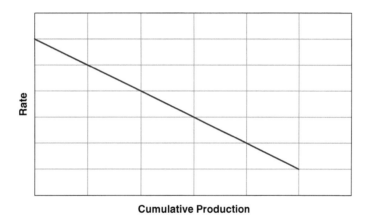

Fig. 1.7—A straight line plot for a rate vs. time curve represents an exponential decline.

Extrapolating a straight line through a $(q$ vs. $Q_p)$ plot results in an estimate of cumulative recovery for the exponential curve **(Fig. 1.7)**.

In conclusion, we can say that straight line semilog rate vs. time and Cartesian rate vs. cumulative production plots define the presence of an exponential decline.

Combine the continuous decline rate and Arps (1945) (*b*-exponent) definition.

$$D = -\frac{1}{q}\frac{dq}{dt} = D_i\left(\frac{q}{q_i}\right)^b \quad \quad (1.24)$$

Arps Hyperbolic Equations When $0 < b \leq 1$. Recall Arps (1945) defined the hyperbolic case to encompass the $(0 < b < 1)$ range and reduced the general rate equation to

$$q_2 = \frac{q_i}{(1+btD_i)^{1/b}}. \quad \quad (1.25)$$

Rearranging Eq. 1.24,

$$t = \frac{\left(\dfrac{q_i}{q_2}\right)^b - 1}{bD_i} \quad \quad (1.26)$$

A rate-decline rate relationship is given by

$$\frac{D_i}{D_2} = \left(\frac{q_i}{q_2}\right)^b \quad \quad (1.27)$$

Substitute the rate equation and integrate; Q_p is the integral of $q(t)$ with respect to t for $(0 < b < 1)$. This results in

$$Q_p = \frac{q_i}{D_i(1-b)}\left[1 - \frac{1}{(1-bD_i t)^{(1-b)/b}}\right]. \quad\quad (1.28a)$$

Substituting the rate equation simplifies to

$$Q_p = \frac{q_i^b}{D_i(1-b)}(q_i^{1-b} - q_2^{1-b}). \quad\quad (1.28b)$$

Assume $(q_2 = 0)$ to express in terms of a theoretical maximum recovery estimate (*EUR*) which is not the EL. Please note that b has to be < 1.

$$Q_{max} = \frac{q_i^b}{D_i(1-b)}(q_i^{1-b}). \quad\quad (1.29)$$

Harmonic Equations. The harmonic case is a restricted version of a hyperbolic case when the exponent term is defined as $(b = 1)$.

The hyperbolic equation reverts to

$$\frac{D_i}{D_1} = \frac{q_i}{q_2}. \quad\quad (1.30)$$

The previously defined harmonic rate equation is

$$q_2 = \frac{q_i}{1+D_i t}. \quad\quad (1.31)$$

Rearrange the harmonic rate equation to determine the time difference spanning two rates.

$$t = \frac{q_1 - q_2}{D_i q_2}. \quad\quad (1.32)$$

To combine rate and time and integrate, use

$$Q_p = \frac{q_i}{D_i}\ln(1+D_i t). \quad\quad (1.33a)$$

Combine with the Arps definition to simplify:

$$Q_p = \frac{q_i}{D_i}\ln\frac{q_i}{q_2}. \quad\quad (1.33b)$$

The Straight-Line Plot. Rewrite the rate equation to a straight-line relationship (**Fig. 1.8**).

$$q_2 = \ln q_2 - \frac{Q_p D_i}{q_i}. \quad\quad (1.34)$$

Straight Line Components

Plotting Variables	Outcome Variables
"y" axes: (ln q)	"y" intercept: (ln q_i)
"x" axis: (Q_p).	slope of the line is: $\left(-\dfrac{D_i}{q_i}\right)$

Table 1.5—Components of Rate vs. Cumulative Production Plot.

For the Arps equations,

- An exponential ($b = 0$) line models single-phase flow from a pressure-depleting reservoir.
- Hyperbolic curves ($0 < b < 1$) model multilayered, gas, or multiphase-flow reservoirs.
- Harmonic ($b = 1$) curves indicate continued presence of transient conditions.

A low value b-exponent indicating eventual decline to a zero rate reflects when boundary flow affects predominate. On the other hand, harmonic and ($b > 1$) values indicate that transient conditions remain and a quantitative reserves estimate is problematical.

Introduction to Decline Curves 11

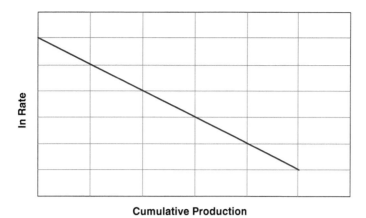

Fig. 1.8—A straight line results when the cumulative recovery equation is rearranged in the form of a straight line.

	Exponential	Hyperbolic	Harmonic
	$b = 0$	$0 < b < 1$	$b = 1$
D	$\dfrac{\ln(q_1/q_2)}{t}$	$D_i\left(\dfrac{q_2}{q_i}\right)^{1/b}$	$D_i\dfrac{q_2}{q_i}$
q	$q_1 \exp(-Dt)$	$\dfrac{q_i}{(1+btD_i)^{1/b}}$	$\dfrac{q_i}{(1+tD_i)}$
Q_p	$\dfrac{q_1 - q_2}{D}$	$Q_p = \dfrac{q_i^b}{D_i(1-b)}(q_i^{1-b} - q_2^{1-b})$	$\dfrac{q_i}{D_i}\ln(1+D_i t)$
t	$\dfrac{\ln(q_1/q_2)}{D}$	$\dfrac{\left(\dfrac{q_1}{q_2}\right)^b - 1}{bD_i}$	$\dfrac{q_1 - q_2}{D_i q_2}$
EUR	$\dfrac{q_1}{D}$	$\dfrac{q_i}{D_i(1-b)}$;	$\dfrac{q_i}{D_i}\ln(1+D_i t)$
		(No q @ EL, so restricted to $0<b<1$).	

Table 1.6—The Arps exponential, hyperbolic, and harmonic rate, time, cumulative production, and decline rate equations.

Bounds of the Arps Equations. Theoretically, the b-exponent term included in the Arps (1945) rate vs. time equation could vary in a positive or negative manner. However, a negative b-exponent value implies an increasing production rate. Therefore, the Arps (1945) equations are truly appropriate only within ($0 < b < 1$) bounds.

Substituting ($b \geq 1$) into the hyperbolic rate equation implies the decline rate is always increasing. This is a nonstarter.

$$q = \frac{q_i}{D_i(1-b)}\left(1 - \frac{q_2}{q_i}\right)^{1-b} \quad\quad\quad\quad\quad\quad\quad\quad\quad\quad\quad\quad\quad\quad\quad\quad (1.35)$$

In conclusion, we can say only exponential and hyperbolic declines converge to zero because the integral of $q(t)\,dt$ is a finite integral (as t goes to infinity for $b < 1$).

When comparing the general semilog rate vs. time plot for the Arps exponential, hyperbolic, harmonic, and $b > 1$ equations:

- The Arps exponential and hyperbolic rate vs. time curves trend in a downward manner to eventually attain a zero rate.
- The harmonic curve does not converge to zero but comes close.
- $b \geq 1$ values do not converge and confirm continuing transient flow generally from a highly variable permeability producing section.

Transient Boundary-Dominated Conditions

An indicator of the drainage radius expanding to boundary dominated flow is the log rate vs. log time plot shown in **Fig. 1.9**. The shape of the log-log plot is a function of the ultimate drainage volume and permeability distribution of the dual porosity system. The transient side usually produces a ½-slope signifying predominantly fracture flow, while the unit slope signifies that the drainage boundary has been reached.

Initializing Decline Curves. Production rates can change because of external and internal factors. When flow occurs after a well or field is temporarily shut in, rates are higher than normal because of the buildup of close-to-wellbore storage. Eventually, production reverts to boundary-dominated conditions after this unsteady state production is unloaded. These effects might impart a segmented curve whose long-term history mirrors the actual depletion history.

Fetkovich (1986) shows in **Fig. 1.10** the effects on the production rate when a North Sea field was periodically shut in. Pronounced production spikes are evident when the closed wells are opened back up. However, the rate soon returns to the normal field decline after inflow has returned to normal rates.

Glenn Pool Field Example. Production history of the Glenn Pool Field in Oklahoma is illustrated in **Table 1.7** and **Fig. 1.11** (Cutler 1924).

Divide the production history into the shaded columns shown in Table 1.7 to compare before and after depletion mechanisms.

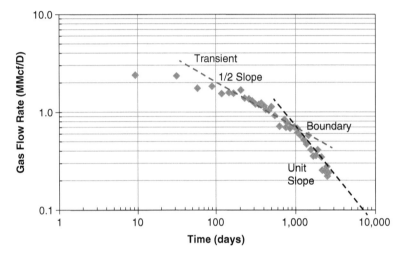

Fig. 1.9—The log rate- log time plot divides well performance history into transient and boundary-dominated depletion regimes. This plot is particularly important when studying low-permeability wells.

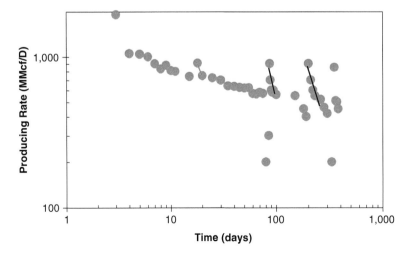

Fig. 1.10—Depletion history for the North Sea field. In each instance production soon declined to the field depletion rate after being reopened to flow. Adapted from Fetkovich (1980).

Time (year)	Initialized Rate	Rate (BOPY)
1	1	10000
2	2	6000
3	3	3400
4	4	2400
5	5	1500
6		1700
7		1850
8		1800
9	1	1750
10	2	1150
11	3	700
12	4	500
13	5	400
14	6	290
15	7	220

Table 1.7—Glenn Pool Field production data.

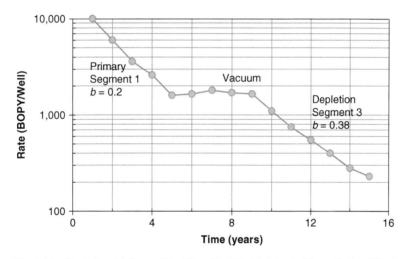

Fig. 1.11—Depletion history of the Glen Pool Field. Adapted from Cutler (1924).

- Segment 1 consists of the initial production decline curve that occurred as the field was depleted to essentially atmospheric pressure.
- Decline ceased at Year 5 when the production system was placed on vacuum and remained essentially constant to Year 9.
- Segment 3 commenced when vacuum operations were discontinued and normal recovery methods were reinstituted. What can we interpret from the production history?

Divide production history into "Primary—Segment 1" and "Depletion—Segment 3" to determine if reservoir depletion reverted to the original mechanism after vacuum operations ceased.

Highlighted values in Table 1.7 reflect two selected production periods. The primary data set was fit to the Arps curve while the Segment 3 set was initialized starting at ($t = 1$) and then fitted to an Arps curve. A good match was obtained for both cases.

Table 1.8 compares the results of analysis. Performance histories of the Glen Pool Field indicated that the vacuum operation produced additional oil. Comparing production history by reinitializing shows that the field has reverted to a hyperbolic decline and that instituting vacuum operations probably accelerated field depletion.

Segment	b	Di	qi
1	0.20	0.64	18,421
2	0.38	0.65	3,170

Table 1.8—Comparison of the decline characteristics for the two segments.

Reserves to Production Ratio. The reserves to production ratio, (R/P) provides a handy screening tool to predict performance when information is scarce.

$$\frac{Reserves}{Production} = \frac{Q_p}{q_{last}} \quad\quad\quad (1.36)$$

Related to the exponential (EUR) equation,

$$\frac{Q_p}{q_{last}} = \frac{1}{D} \quad\quad\quad (1.37)$$

Please note that this is EUR and not EL.

The value provides a useful screening tool to evaluate the well quality. Most of the wells should cluster in the middle of the plot, but good and bad wells that should require further evaluation are located at the ends of the spectrum.

Reserves are calculated by decline curve analysis or by some other means.

PROBLEM(S)

Example Problem 1.1. Apply exponential decline curve analysis techniques to analyze the Wafford No. 1 well rate vs. time history. See **Fig. P.1.1.1.**

Learning Objectives. Realize applying either the nominal or the effective exponential decline equations results in similar answers, and understand fitting a line to production data is often an individual selection process. Please answer the following questions.

1. Draw a best fit straight-line approximation of the performance history.
2. Determine effective decline rate. Compare decline rate over a 1-year period.
3. Determine the nominal decline rate.
4. Compare the two answers.
5. Calculate expected producing rate at Month 28?
6. How much longer will the well produce when the economic limit is 10 BOPM?
7. How much oil will be produced between Month 28 and the 10-BOPM economic limit?

Example Problem 1.2. Apply the exponential concept to calculate the effect of well clean out on performance.

Learning Objective. Apply exponential rate and cumulative production equations to evaluate reserves potential for working over the Hollands No. 3A well.

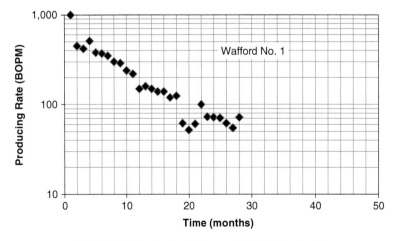

Fig. P1.1.1—Producing history of Wafford No. 1 well.

The Hollands No. 3A well currently displays a $D = 37\%/\text{yr}$ decline rate and produces at 52 BOPD. Replacing the pump and scraping the producing string would increase the rate from 52 to 96 BOPD but not change the reserves picture. This is a rate acceleration problem. Economic Limit = 6 BOPD.

Compare a "Do Nothing" case to the "Remedial" case to calculate the economics of the projected workover expense.

Useful equations: $N_p = \dfrac{q_1 - q_2}{D}$, $\quad D = \dfrac{\ln(q_1/q_2)}{t}$.

Hints:

- The endpoint for the "Do Nothing" and "Remedial" cases is the volume of oil that could theoretically be produced to the estimated ultimate recovery (EUR).
- Calculate (EUR) by assuming $q_{\text{last}} = 0$.

Set cumulative production for the "Do Nothing" and "Remedial" cases equal to each other. Calculate the decline rate for the "Remedial Case", as shown in Table 1.9.

- Apply the system of equal triangles to determine the new decline rate.
- Calculate a new rate vs. time forecast to compare with the old forecast.

Year	Do Nothing		Remedial		Incremental
	Rate (BOPM)	Cum. (BO)	Rate (BOPM)	Cum. (BO)	(BO)
0	52		96		
1	36	15,784	49	25,117	9,333

Table 1.9—Comparison table.

Calculate and plot the rate vs. time schedule on the following graph.

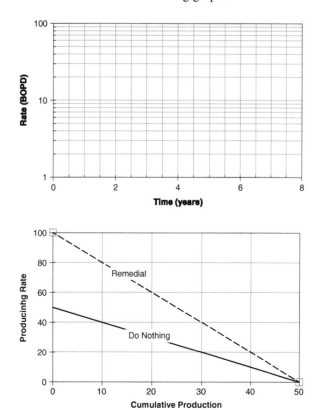

Fig. P1.2.1—Producing history of Well #3A.

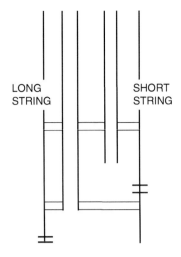

Fig. P1.3.1—Completion setup of a dually completed gas well.

Example Problem 1.3. Diagnosing a well problem.

Learning Objective. A study of production records can aid in interpretation of source of excess water production.

A gas well was dually completed in two pay zones at approximately 5,500 ft: generally, sands in this area are friable. In fact, both completions do produce some sand. **Fig. P1.3.1** shows the completion setup with the lower (long string) and upper (short string) set of perforations separated by a packer.

Recently water production has been observed in the long string, and the rate has declined to approach that of the upper sand.

Is this effect caused by normal water encroachment in the reservoir or by a hole eroded into the blast joint?

Interpret performance histories of a dually completed well to find the unwanted water source as well as obtaining insight into reservoir performance.

Fig. P1.3.2 compares gas production rates for the two zones. Note the erratic gas production rate from the short string.

Fig. P1.3.3 shows the track of the flowing-tubinghead pressure (FTHP) for the two completion zones. Water production was consistently measured.

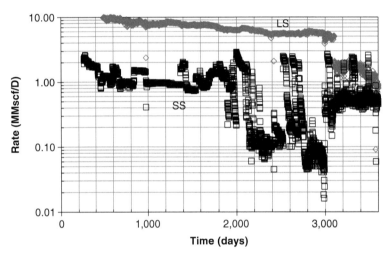

Fig. P1.3.2—Notice the long string (LS) produced at a much higher rate than the short string (SS). Decline rate for the deeper well, $D = 5\%/yr$. The short string experienced sanding problems over much of its producing life. Well problems caused erratic production rates after approximately 1,900 days.

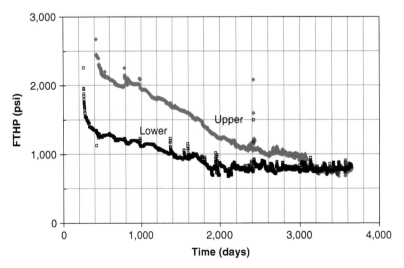

Fig. P1.3.3—Flowing-tubinghead-pressure history for the two completions.

Consider these questions:

- What is your interpretation of the histories of the two completions?
- Does water encroachment affect reservoir performance?
- Can you estimate when communication between the two production strings began?

Example Problem 1.4. Has the well watered out or is there a hole in the tubing?

Learning Objective. Couple well locations with performance analysis to determine source of unforeseen water production.

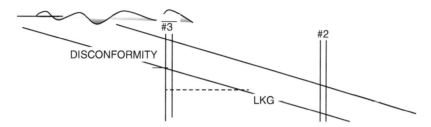

Fig. P1.4.1—Well locations and structural interpretation.

Fig. P1.4.1 shows that the #2 well was drilled in a downdip location (defined by an unconformity) in a layered, friable sand. A lowest known gas (LKG) was observed. Perforations were at the top of the sand. However the #2 well recently watered out.

One year later, the updip #3 well was drilled and encountered a gas-filled sand similar in nature to that of #2. The well was completed and inflow performance 34 MMscf/D with no water.

History shown in the **Fig. P1.4.2** indicates that the #2 well produced at 20 MMscf/D until 1,000 days when it sanded up and was down for approximately 600 days. The well never returned to its initial potential after gravel pack.

On day 1,800, the #2 well began to cut water and 2.2 years later was shut in because of high water production.

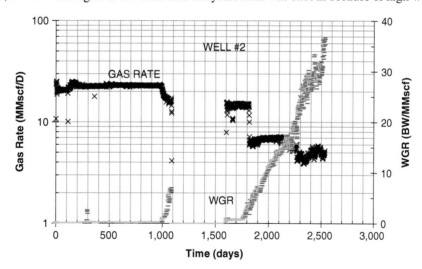

Fig. P1.4.2—Producing history of Well #2.

The updip #3 well continues to produce at approximately 20 MMscf/D essentially water free. (**Fig. P1.4.3**).

Example Problem 1.5. A north texas gas condensate well.
  Fig. P1.5.1 shows the 5-year history for the gas-condensate well.

Learning Objectives. Relate changes in field operations to changes in the shape of the decline curve.
Please answer the following questions.

1. Two workovers occurred during the life of the well. Can you spot the probable time of these workovers? Were they effective?

2. What can you say about the consistency of the gas and condensate producing rates?
3. What eventually killed the well?
4. Was there a hole in the tubing?
5. What is the probable cause for the increased water production early in the well life?

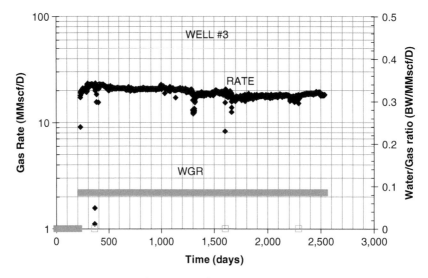

Fig. P1.4.3—History of well #3.

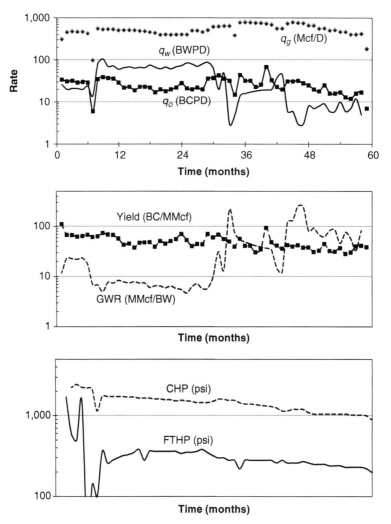

Fig. P1.5.1—A north Texas gas/condensate well. CHP = Casinghead pressure.

Chapter 2
Effects of Field Conditions

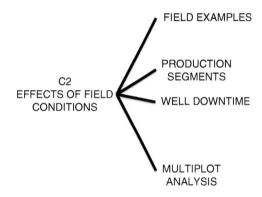

Analysis of well performance assumes an unobstructed flow and constant operating conditions. Any downhole and surface changes in the flow system are usually reflected in the production rate, which in turn transforms performance into operating segments. Several examples will illustrate the effects of different conditions.

The well drainage system depends on pressure loss across the drainage sections to overcome the hydrostatic head resisting flow up the wellbore. Therefore, reservoir and operational characteristics affect the shape of the performance decline curve.

Knowledge of reservoir rock and fluid type and expected drive mechanism can be of considerable aid when interpreting a production decline history.

Last, simultaneously studying multiple different-scale performance plots can be of considerable aid when interpreting and predicting future character.

Candidates lending themselves to decline curve analysis are

- Wells suffering a measurable and continuing loss in reservoir pressure.
- Wells without mechanical restrictions and adequate communication with the producing interval.
- Gravity-drainage reservoirs (implied high permeability).
- Reservoirs where the mobility-thickness product $\left(\frac{kh}{\mu}\right)$ is sufficiently high.
- Reservoirs not connected to waterdrive, gas-cap expansion, or overpressured additive pressure influence.
- Wells producing essentially a single-phase fluid. This statement is not always valid, but individual-phase production rates should be analyzed.

Fig. 2.1 shows a schematic of the flow system from the reservoir to the well (inflow) and the flow from the bottom of the well to surface (outflow). On the basis of a simple production system, the total pressure drop is the sum of the pressure drops in the reservoir, vertical tubing and the horizontal pipe, plus all the pressure drop resulting from accessories at the surface (choke, flowline, separator). Each pressure drop depends on flow rate.

It can be observed that fluids flow in the reservoir between $(r_e \to r_w) \to$ hydrostatic head between $(p_{wf} \to p_{th}) \to$ pressure drop through surface facilities.

This chapter shows examples of the effects of changing surface and subsurface conditions on pressures and flowing conditions at the different nodes in the system. Those possible changes will affect reservoir and well

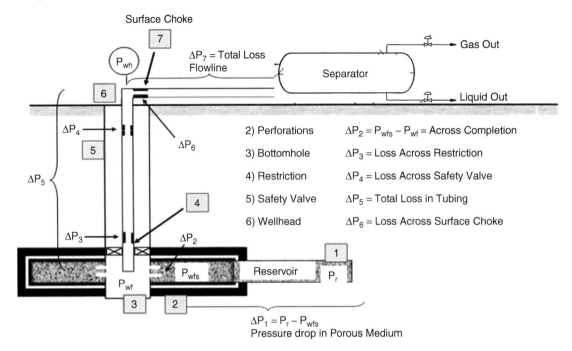

Fig. 2.1—Production System schematic. Flow rate is affected by total pressure drop across the nodes of the integrated system (not showing all possible surface accessories). [Adapted from *Reservoir Stimulation*; Economides and Nolte; 3rd Edition (2000)].

analysis when predicting producing rates. Important questions normally raised concerning those possible changes include these: What fluids will it produce? For how long? Under what uncertainties?

Field Examples

Fig. 2.2 shows the effect on an Ellenberger gas well when the line pressure was increased to 600 psi (Huddleston 1991). A well-defined exponential decline occurred over the first 9.5 years of life. Increased backpressure significantly altered the reserves estimate, and the decline rate increased from 17%/yr to 26%/yr.

Reserves can be overestimated if a change in external conditions is not taken into account. **Fig. 2.3** provides an example of the effect of water influx on maintaining a well-defined, 20%/yr decline rate. The gas rate and water/gas ratio (WGR) remained stable for approximately 34 months. Extrapolation of the obvious straight line into the future would overestimate reserves. The decline between month 34 to 49 was caused by the well gradually loading up with water and increasing pressure-drop requirements between the perforations and tubing head.

Decline curve analysis should not be performed on tight reservoirs in the early stages of depletion unless one understands that transient conditions control performance. An extended prediction tail will occur only after boundary-dominated flow conditions eventually dominate. Early time data are a reflection of drainage close to the wellbore when the apparent reservoir limit is small.

Fig. 2.4 represents the production history of a dual-porosity Austin Chalk well. The Austin is composed of a very-low-permeability calcareous mudstone transected by fractures running parallel to the Gulf Coast basin.

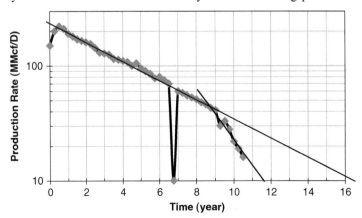

Fig. 2.2—Effect of increasing line pressure. How could you calculate decrease of ultimate recovery? (Adapted from Poston and Poe 2008).

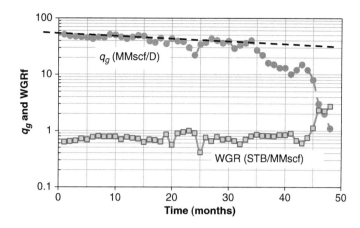

Fig. 2.3—Effect of a waterdrive on the performance history of a gas reservoir. Adapted from Huddleston (1991).

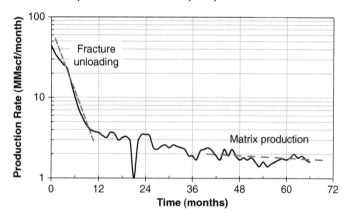

Fig. 2.4—A dual-porosity Austin Chalk well. (Adapted from Poston and Poe 2008).

Extrapolation of the early decline to a minimum rate indicates that the well would last for only 14 months. The second line indicates that the well should produce as long as the rate remains above the economic limit and mechanical problems do not arise.

More-complex analysis techniques for predicting performance are discussed in a later chapter.

Production Segments

Fig. 2.5 shows the production history of a waterflooded Canadian reservoir, (Baker et al. 2003). For this case, one would expect the history could be divided into at least a primary period and waterflood period. However, one must be aware that an external energy source is now affecting recovery.

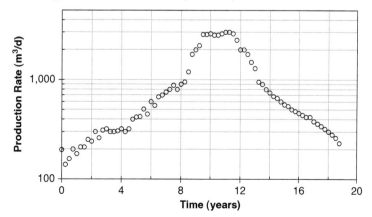

Fig. 2.5—History of Canadian field lasting from development to production plateau to water breakthrough and gradual loss of oil production rate. (Adapted from Baker et al. 2003).

The history can be divided into a series of events, which are

- Additional field development and water injection commenced in the third year.
- Water injection caused a dramatic increase in production when selected infill drilling occurred to optimize with the full effect seen in Year 9.
- Production was flat until Year 11 when significant water breakthrough commenced.

How would you forecast future production? How would you predict future performance? The decline portion of the history shown in Fig. 2.5 was fitted to a semilog rate. **Fig. 2.6** shows the time plot of three graphs, which compare matches over different portions of the production history. Which do you think is appropriate?

What can one say about applying regular decline curve analysis methods to study performance? Generally, Arps methods of applying decline curves to predict future performance in the presence of any type of injection should be avoided. One can see that predicting future performance by simple decline curve analyses techniques

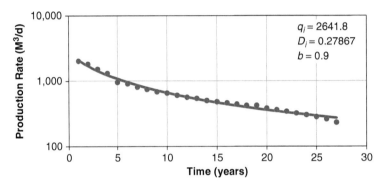

The first match extends over the entire decline of Segment 3. The red line shows the fit on the curve.

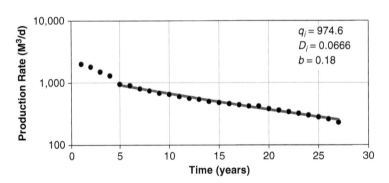

The second match extends after the initial steep decline.

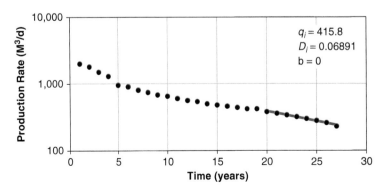

The last match extends over the last few data points when there is a pronounced slope change. Results of these analyses are presented in **Table 2.1.**

Fig. 2.6—Comparing Matches.

	History		
1 → 27	$b = 0.93$	$q_i = 2{,}642$ m³/d	$D_i = 0.28$/yr
5 → 26	$b = 0.18$	$q_i = 975$ m³/d	$D_i = 0.061$/yr
20 → 27	$b = 0.0$	$q_i = 416$ m³/d	$D_i = 0.069$/yr

Table 2.1—Comparison of Predictions.

can entail uncertainties if one does not apply all facets of knowledge when analyzing the problem. Both geological interpretation and reservoir behavior must be taken into account.

Example. Changing operating conditions can alter a well history to such a degree that long-term trends in producing character appear indecipherable. **Fig. 2.7** shows the history of a well located in a Colombian field.

A series of line approximations representing individual correlation segments, of oil producing rate, gas/oil ratio (GOR), and water cut can be drawn for each period. Therefore, an apparently highly erratic history is changed to a series of reasonable approximations reflecting operations occurring at that time. The smoothing process divided the well history into three periods and one period of no information.

Correcting for Well Downtime

A well is not necessarily operated or produced for every day of the month. Mechanical problems or bad weather might cause a well to be temporarily shut in.

Production records saved at a governmental body are generally displayed on a monthly total basis. Dividing monthly cumulative production by the number of "calendar days" does not necessarily result in a true average production rate if the well did not produce for each day of the month.

The true average production rate is reflected only by dividing the number of "operating days" into production since the number reflects the actual number of days the well was produced.

A well is defined as "underutilized" when the well did not produce each day of the month.

$$Underutilized,\ calendar\ day\ rate = \frac{Production\ over\ calendar\ period}{Days\ in\ calendar\ period} \quad\quad (2.1)$$

The producing or actual day rate divides totaled production accumulated during a particular time period by the number of days the well was actually producing. This rate reflects the maximum potential if production is not being curtailed.

$$Fully\ utilize,\ producing\ day\ rate = \frac{Cumulative\ production}{Number\ of\ days\ on} \quad\quad (2.2)$$

Example. Comparing the producing record of a Mannville oil well. Adapted from Purvis (1984).

Fig. 2.8 shows that a Canadian well did not produce each day of the month until Year 5.3 when the rate became fully utilized and produced continuously. The producing GOR shown in **Fig. 2.9** inexplicably more than doubled

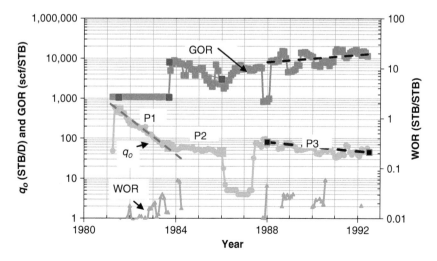

Fig. 2.7—The 10-year history of the L95 well. The history could be divided into at least three periods when field conditions appear to remain constant. WOR = water/oil ratio. Poston and Poe (2008).

in Year 5.3, and reservoir drawdown increased. **Fig. 2.10** shows an exponential rate decline after production rate capacity was fully utilized. In actuality, gas production went up as a response to the increased drawdown.

Informational Plots

Fig. 2.11 represents a plot of well producing rates for nine tracts located in the Salt Creek Field, Wyoming (Cutler 1921). In this case, production is derived from homogeneous, high-permeability sand that appears continuous over the field.

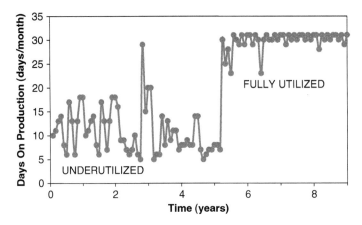

Fig. 2.8—Well produced at monthly capacity only after 5.3 years of life. Average producing rate calculations would obviously differ. (Poston and Poe 2010).

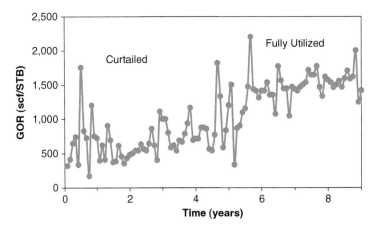

Fig. 2.9—Is the increased producing GOR a product of increased drawdown close to the wellbore because of a higher production rate? Analysis of reservoir performance would be quite dissimilar (Poe and Poston 2010).

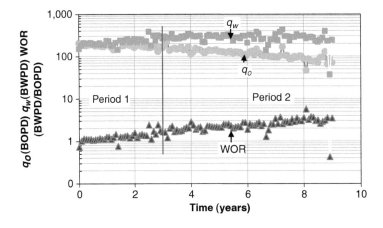

Fig. 2.10—Water rate remains remarkably constant while oil rate declines at 20%/yr when fully utilized (Poe and Poston 2010).

Tract production was divided by the number of wells producing from the property, to calculate an average per-well producing rate.

Well Records. Well tests include oil, gas, and water production rates; flowing and shut-in pressures; and oil-gravity measurements. A plot of these data serves as a reference to analyze total well behavior. **Fig. 2.12** is a record of the, B-8T well, offshore Texas.

This discussion illustrates the amount of information available when all performance data are plotted on a single figure.

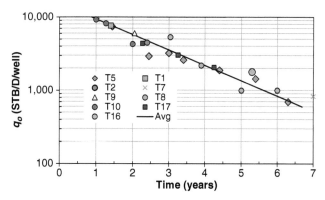

Fig. 2.11—History of the Salt Creek Field. The straight line correlation indicates wells are generally in communication. Estimating future reserves can be achieved with confidence. Adapted from Cutler (1924).

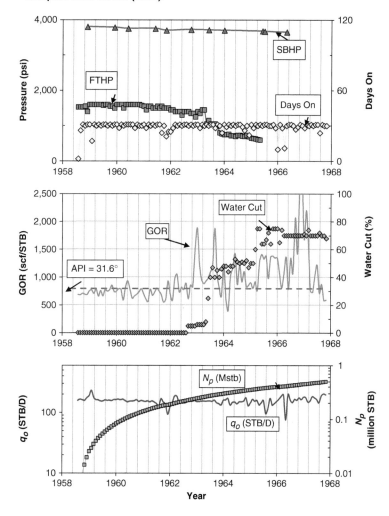

Fig. 2.12—Various reservoir parameter histories placed on the same time scale for interpretation. Adapted from Huddleston (1991).

Analysis. Abrupt change in flowing tubinghead pressure (FTHP) was probably caused by water production increasing the hydrostatic head in early 1963. There has been little effect on the oil producing rate even though there was a significant increase in water production. The only minor decrease of the shut-in bottomhole pressure (SBHP) indicates an effective waterdrive. Which records indicate the onset of water production?

The API gravity of the crude did not change significantly. Dramatic changes in the API gravity often indicate the presence of wellbore communication with multiple pay sands completed in a wellbore.

Multiplot Analysis

The majority of current decline curve analysis is conducted with a single (log rate vs. time) plot. In reality, more information concerning well and reservoir character can be obtained when multiple plots representing different coordinate systems are constructed and analyzed.

The following proposed problems shows that simultaneously studying performance histories plotted in different coordinate systems can broaden the scope of a study considerably. Apply these plots to interpret well or field characteristics. Each figure tells only a portion of the story, but coalescing disparate plots into the study furnishes a more-complete picture of the performance characteristics.

Listed below are performance plots used in the problems, but other plots could be included according to particular flow systems.

- Semilog rate vs. time
- Rate vs. cumulative production
- Logarithmic rate vs. Logarithmic time

PROBLEMS

Example Problem 2.1. Analysis of performance history for a Gulf of Mexico (GOM) field.

Learning Objective. The learning objective is to perform calculations that show "fully utilized" and "calendar days" definitions that might result in different answers. See **Fig. P.2.1.1** and **Table P.2.1.1**.

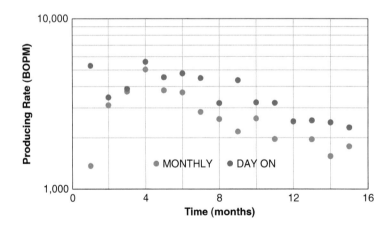

Fig. P2.1.1—Producing history of well.

Month	Calendar (days)	Fully Utilized (days)	Month (BOPM)	Month (BOPD)	Days On (BOPD)
03/07	31	8	42,365	1,367	5,296
04/07	30	27	93,256	3,109	3,454
05/07	31	30	115,989	3,742	3,866
06/07	30	27	150,998	5,033	5,033
07/07	31	26	117,791	3,800	4,530
08/07	24	31	114,555	3,695	4,773

Table P2.1.1—Production Data, GOM well.

Answer the following questions.

1. Verify the "fully utilized" and "calendar" rates expressed in BOPD.
2. Draw straight line approximations through each data set, and calculate the decline rate of each.
3. Which definition, "fully utilized" or "calendar" rates expressed in BOPD, results in the most reliable answer?
4. Calculate reserves if the economic limit is 1,000 BOPD.

Example Problem 2.2. Performance history example of a west Texas hydraulically fractured oil well.

Learning Objective. Review a well history to illustrate how much information can be developed from raw performance plots.

Interpret well and depletion history.

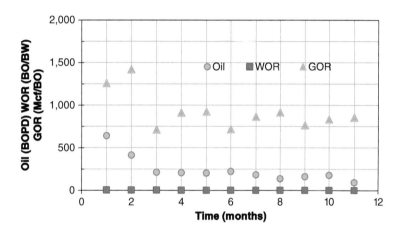

Fig. P2.2.1—Cartesian plot that shows the historical oil producing rate, water/oil ratio, and gas/oil ratio.

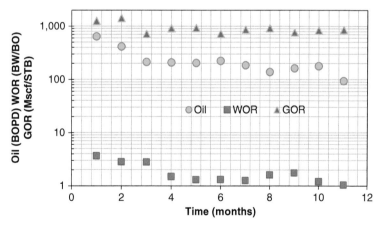

Fig. 2.15—Semilog performance plot of Fig. P2.2.1.

Answer the following questions.

1. Establish the number of producing segments. Determine the time interval over which each segment predominates.
2. Calculate the decline rate over the apparent time interval over which boundary dominated flow predominates.
3. What does the essentially constant GOR reflect about pressure history?
4. What can we say about the nearly constant WOR?

Example Problem 2.3. An Ellenberger gas well.

The semilog rate vs. time plot in **Fig. 2.16** displays a well-defined exponential decline over the first 9.5 years of life at the time when line pressure was increased to 1,800 psi from 600 psi.

Learning Objective.

- Review a well history to illustrate how changes in operation conditions will affect performance and recovery.

Please calculate the loss of production.

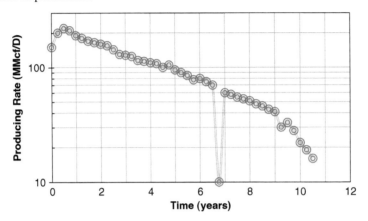

Fig. 2.16—Performance history for the Ellenberger well.

Example Problem 2.4. The Bollycotton gas field was discovered offshore Ireland. **Table P2.3.1** lists reservoir properties and OGIP listed in the Chu et al (2001).

Learning Objective. Review a field history to illustrate and apply the Arps and quadratic concepts to analyze production history.

Reservoir Rock and Fluids Properties				
p_i = 1,200 psi	γ_g = 0.554	h = 76 ft	G_p = 25.3 Bscf	OGIP = 53.2 Bscf
T_{res} = 120 °F	T_{res} = 580 °R	ϕ = 22.3%	k = 100 to 109 md	μ_g = 0.013 cp

Table P2.3.1—Data for Example Problem 2.4, Bollycotton Field.

Compare bottomhole flowing pressure and rate decline histories. The apparent lack of significant water production and considerable pressure loss seem to indicate pressure depletion conditions. Performance history is shown in **Fig. 2.17**.

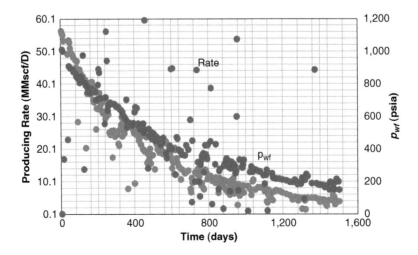

Fig. 2.17—Production rate and bottomhole flowing pressure dramatically of four years production history. Evidence of little or no water production is very low flowing bottom hole pressure.

The semilog rate vs. time history (**Fig. 2.18**) seems to represent an exponential decline?

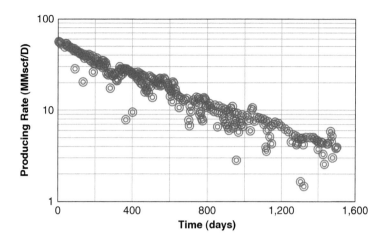

Fig. 2.18—The Bollycotton field performance history.

Determine decline rate (D = fraction/year) and initial rate (q_i)

Summarize your findings.

Chapter 3

Smoothing Variable Production

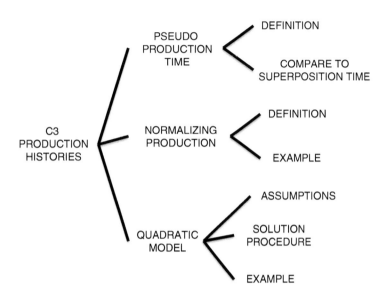

PRODUCTION HISTORIES

The following discussion develops ideas about editing and upgrading field oil, gas, and water production data. Subjects discussed are

- Pseudo production time (t_{mb} or t_p) calculations typically smooth producing rate fluctuations to form a more uniform curve. The smoothing process often permits interpretation of performance with a higher degree of certainty.
- Computer programs are available for converting flowing tubinghead pressures (p_{tf}) to bottomhole flowing pressures (p_{wf}). The calculating procedure accounts for functionalities that consider the effect of fluids properties, multiphase flow correlations, friction factor, tubing diameter, and other factors on producing rate. Normalizing the rate by dividing by the changing pressure differential often produces a much smoother production curve.
- The quadratic equation plots (p_{wf}/z) vs. cumulative gas produced to determine original gas in place. The technique obviates the need for static pressures (p_{res}) in the well-known (p/z) plot. A simplifying assumption has reduced the complexity of the solution procedure.

Pseudo Production Time

Horner (1951) demonstrated that an equivalent material balance time (t_p) was useful when calculating production decline behavior for the variable-flow-rate case. Palacio and Blasingame (1993) included the material balance time (t_{mb}) concept that is equivalent to the venerable Horner time definition in their formative work. These definitions were coalesced to a single definition "pseudo production time (t_p)" for the following discussions.

$$t_p = \frac{N_{p(STB)}}{q_o\left(\frac{STB}{day}\right)}, \text{ in days.} \tag{3.1}$$

The approximation is intuitive, does not have a theoretical basis, and works best when rate changes are small and short-lived. Many investigators applied this approximation because of the simplicity of the calculation procedure and its smoothing abilities.

Question: How do you handle two or more production interruptions?

Example. A well has produced 273,521 BO and recently tested an average 2,320 BOPD for the last few months. Calculate material balance time.

$$t_{mb} = t_p = \frac{N_p}{q_{last}} = \frac{273,521\,\text{BO}}{2,320\,\text{BOPD}} = 117.9\,\text{days} = 117.9\,\text{days}.$$

Express dimensionless material balance time by two different equations.

$$t_{Dp} = \frac{Q_{pD}}{q_D} \quad \text{...} \quad (3.2)$$

Poe and Marhaendrajana (2002) and Poe and Poston (2010) redefined dimensionless material balance time as

$$t_{Dp} = \frac{kt_{mb}}{\varnothing \mu c_t L_C^2}. \quad \text{...} \quad (3.3)$$

Note that the substitution of the characteristic length (L_c) for (r_w) in the denominator of Eq. 3.3 is equal to one-half the lateral extent of the well source/sink. Eq. 3.3 encompasses all possible completion configurations, as shown in the following.

For instance,

- $L_c = (r_w)$ for an unfractured vertical well.
- L_c = the fracture half-length for a fractured well $\left(\frac{X_f}{2}\right)$
- L_c = half of the effective wellbore length (L_D) for a horizontal well $\left(\frac{L_h}{2}\right)$.

Compare Pseudo Production Time (t_p) to Superposition Dimensionless Time (t_d).
van Everdingen and Hurst (1949) applied superposition dimensionless time (t_D) to account for rate or pressure fluctuations to develop solutions to the diffusivity equation.

Applying pseudo production time helps smooth production fluctuations. However, a substitution of pseudo production time to replace the superposition function in certain flow situations can result in significant error for low-permeability cases. Refer to: Poe and Marhaendrahana (2002).

Recall the comparison of real time (t) with dimensionless time (t_D) comparison:

$$t_D = \frac{0.00633kt}{\varnothing \mu c_t r_w^2}, \text{ in days} \quad \text{...} \quad (3.4)$$

The (t_{Dp}/t_D) ratio reduces to the following when defined in dimensionless terms:

$$\frac{t_{Dp}}{t_D} = \frac{t_{mb} r_w^2}{0.00633\, t\, L_C^2}. \quad \text{..} \quad (3.5)$$

Poe and Mahaendrahana (2002) showed that the material balance and superposition dimensionless time functions for early rate-transient performance of a well can differ by as much as 200% during the pseudolinear (or formation linear) flow regime. These differences are illustrated in the next three figures when dimensionless time is plotted as a function of (t_{Dp}/t_D) ratio.

Unfractured Well. **Fig. 3.1** shows the effects of skin on the (t_{Db}/t_D) ratio for an unfractured well producing from an infinite acting reservoir. The only case in which the two time expressions coincide throughout the whole producing life is the case of a heavily damaged well. Otherwise, pseudo production time (t_p) may be off by 200% in the regions where it really counts.

Fractured Vertical Well. A fractured vertical well producing from an infinite acting reservoir showing the effects of the dimensionless fracture conductivity (C_{fD}) expression on the (t_{Dp}/t_D) ratio is shown in **Fig. 3.2**.

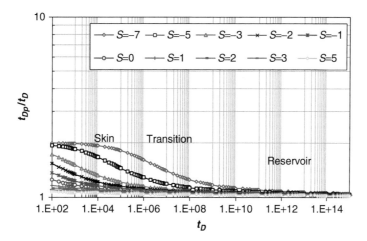

Fig. 3.1—Higher than expected matrix stimulated permeability causes an extraordinarily high (t_{Dp}/t_D) ratio initially. Eventually, production reverts to matrix depletion conditions. For instance, an acid wash induces an artificially high permeability to increase rate at early time. Adapted from Poe and Marhaendrajana (2002).

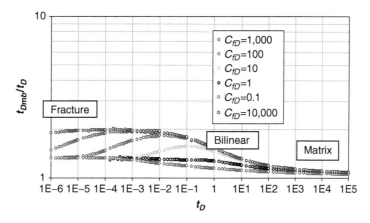

Fig. 3.2—Early flow conditions of a vertical, fractured well. A highly conductive fracture causes the balance time to dramatically diverge from the correct time frame (t_D). Once again, we see that error might be as much as 200% for high-fracture-conductivity wells, particularly at early and middle times. Lag time increases as the fracture system becomes less extensive and less efficient. Adapted from Poe and Marhaendrajana (2002).

Recall that: $C_{fD} = \dfrac{k_f b_f}{k X_f}$. .... (3.6)

The fracture flow capacity term is a measure of how easily fluid moves through a fracture in relation to the matrix contribution.

Horizontal Wellbore Length. The effect of dimensionless wellbore length on the (t_{Dp}/t_D) ratio for a horizontal well producing in an infinite acting reservoir is shown in **Fig. 3.3**.

Recall that, $L_D = \dfrac{L_C}{h}\sqrt{\dfrac{k_z}{k}}$. .... (3.7)

Dimensionless wellbore capacity (L_D) relates flow in the horizontal completion string and compares how easily fluid flows in the vertical direction to the average permeability of the system as a whole.

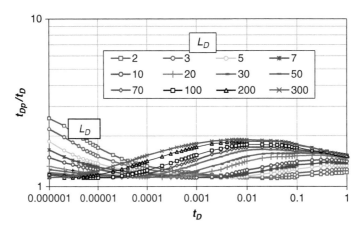

Fig. 3.3—The extended-horizontal-wellbore case. Adapted from Poe and Marhaendrahana (2002).

The ratio could be off by more than 200% for the short-horizontal-well case. Notice how bilinear flow for the longer-wellbore case causes the ratio to increase later in well life. Fig. 3.3 shows how the storage capacity of the horizontal component will affect the well flowing character. This is shown when

- The shortest wellbore length, $L_D = 2$, rapidly reverts to the matrix contribution after unloading the fluid that was initially in storage state at the wellbore.
- Wellbore storage extends the occurrence of the maximum rate when ($L_D = 300$).

The discussion demonstrated that the pseudo-production-time function will always be greater than or equal to the corresponding superposition time function at a given time level. Therefore,

- The ratio of pseudo production to superposition time function $\left(\dfrac{t_{Dp}}{t_D}\right) > 200\%$ during the pseudolinear flow regime exhibited during the transient performance of moderate to high-dimensionless-conductivity fracture systems. However, the two values will coincide at a later time.
- The relationship between pseudo production and superposition time generally varies continuously during early transient behavior of low-permeability sands.
- Variations between flow rate and pressure drop are most exaggerated when producing from low-permeability reservoirs.
- Incorrect usage of pseudo production time for the low-permeability case can cause significant errors in rate vs. time calculations.
- The superposition function compensates for these inaccuracies, but is somewhat complicated to apply.

Normalizing Production

Apparently, Miller (1942) was the first to recognize that monitoring bottomhole flowing pressure (p_{wf}) changes could aid when predicting performance. Normalizing production rate—that is dividing it by pressure drawdown—can smooth seemingly erratic production rates and reveal long-term trends. Eq. 3.8 presents the relationship.

$$q_{nor} = \left(\frac{q}{p_i - p_{wf}}\right). \quad\quad\quad\quad\quad\quad\quad\quad\quad\quad\quad\quad\quad\quad\quad\quad\quad\quad\quad\quad\quad\quad (3.8)$$

Eq. 3.8 works only for the pressure-depletion case. Addition or deletion of any external energy source in the producing system alters history.

Flowing tubinghead pressures (p_{tf}) are often measured during a well production test. Vertical profile programs are available for calculating differences between flowing top-hole pressure (p_{tf}) and bottomhole pressure (p_{wf}). This information can be applied when flowing pressures are available.

Example. **Table 3.1** lists a portion of a spreadsheet and work detail from Blasingame and Lee (1986), where $p_i = 2{,}000$ psi.

Time	Rate	1 Cumulative Production	2 p_{wf}	3 t_p	4 t_D	5 $q/\Delta p$
(days)	(STB/D)	(STB)	(psia)	(days)		(STBPD/psi)
30	1,500	45,000	1,607	30	5.06×10^7	3.82
60	700	66,000	1,800	94	1.01×10^8	3.50
90	1,900	123,000	1,480	65	1.52×10^8	3.65

Table 3.1—Rate, time, pressure data from Blasingame and Lee (1986) Example Well 1.

Fig. 3.4 shows a very erratic flow rate and conversely bottomhole flowing pressure (p_{wf}) for the reference well. In fact, there is no obvious trend line.

Fig. 3.5 shows the same information in normalized form. Columns 5 in Table 3.1 lists the calculated values. Compare with Fig. 3.4.

The Quadratic Model

Another way to apply flowing tubinghead pressures (p_{tf}) to calculate reserves for a gas well is to apply the quadratic model, which accounts for changing fluid properties at low pressures. There is no need for static bottomhole pressures as is the case for the (p/z vs. G_p) plot.

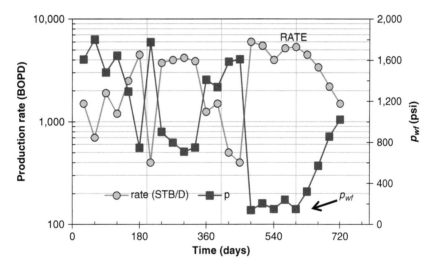

Fig. 3.4—Raw rate vs. time data from the Blasingame and Lee (1986) example. Adapted from Blasingame and Lee (1986).

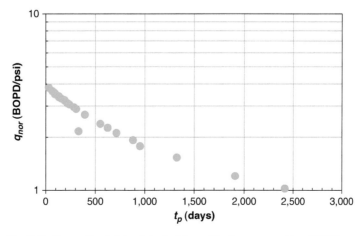

Fig. 3.5—Normalized data. Adapted from Blasingame and Lee (1986). Note the smoothly declining curve. Compare the decline to that of Fig. 3.4.

Normally, one extends a straight line downward to estimate reserves on the semilog rate vs. time plot for a pressure-depletion gas reservoir, as shown in **Fig. 3.6**. The volume could be underestimated in pressure depleting, low-to moderate-pressure gas wells because of changes in (z-factor and μ_g) at low pressures.

The effects of pressure on gas viscosity (μ), z-factor, and the (μ_z) product term are shown in **Fig. 3.7.**

Conclusions. Changing fluid properties particularly at very low pressures can affect the semi-log rate vs. time plot.

Fraim and Wattenbarger (1987) showed that a gas well producing from a pressure-depleting reservoir does not necessarily match the exponential curve at low pressures.

Experience has also shown that a very-low-pressure, long-lived gas reservoir production decline curve can be modeled equally well with an exponential or a harmonic decline curve. Normal decline theory holds that this should not happen.

Knowles (1996) and Ansah et al. (1996) developed a quadratic relationship for the (producing rate vs. cumulative recovery) plot.

Differences between the shapes of the exponential and quadratic (q_g vs. G_p) curves are compared in **Fig. 3.8.**

For an exponential decline, the Arps exponential straight line begins at q_{gi} and it ends at $q_g = 0$, i.e., ($G_{p(\max)}$).

For a quadratic decline, the equivalent quadratic equation line is initiated at q_{gi} and extends in a linear fashion until trending in a concave upward manner, with the curve ending below the $q_g = 0$ value.

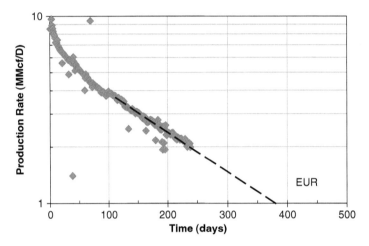

Fig. 3.6—Semilog rate vs. time plot. One could underestimate gas reserves because of changing fluid properties at very low pressures with the rate vs. time plot.

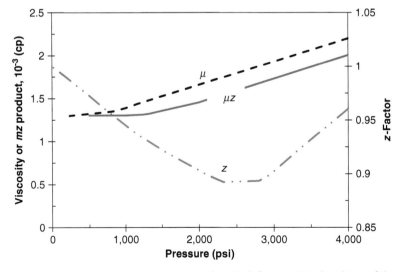

Fig. 3.7—Note, the (μz) product curve is primarily influenced by the shape of the *z-factor*, particularly at pressures < 1,600 psi.

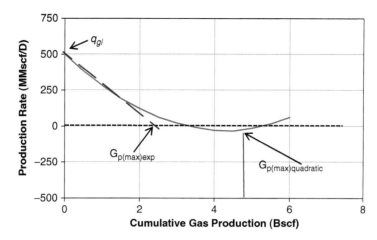

Fig. 3.8—Comparing the exponential curves. The quadratic expression accounts for the effect of gas properties at low pressures. Adapted from Knowles (1996).

The minimum point is the "quadratic root" and is equivalent to an estimated ultimate recovery (EUR). The physical meaning for the producing rate trending below zero for the quadratic equation is not clear. Adapted from Buba (2003).

Define the quadratic curve in terms of q_g, G_p, and p_{wf}:

$$q_g = q_{gi} - \frac{2q_{gi}}{\left(1 - \frac{p_{wf}/z_{wf}}{p_i/z_i}\right)G} G_P + \frac{q_{gi}}{\left[1 - \frac{p_{wf}/z_{wf}}{p_i/z_i}\right]G^2} G_P^2 . \quad\quad (3.9)$$

Four basic parameters are required to solve the equation:

- Initial pressure (p_i)
- Flowing pressure history (p_{wf})
- z-factor
- Initial producing rate (q_i)

Note the absence of the static reservoir pressure (p_{avg}) history parameter in the calculation procedure. A bottomhole flowing pressure (p_{wf}) history is the only requirement. Jordan et al. (2009) assumed that q_i may be determined if an exponential curve is present.

Assumptions and Limitations. The assumptions and limitations are

- Pressure depletion conditions and $p_i <$ 6,000 psi.
- Assume negligible rock and water compressibility and capillary pressure affects.
- The original gas in place (OGIP) value is predicated on the validity of the exponential extrapolation to determine q_i.
- Adding cumulative recovery (G_p) back into the solution of Eq. 3.14 results in a horizontal, constant line on the G_a vs. G_p during the steady state depletion history.

Including the Jordan et al. (2009) idea of finding initial flow rate with decline curve analysis changed an intractable problem to a practical oilfield solution. This method is very useful because a flowing bottomhole pressure history can be substituted for a static pressure record. One must be aware that the q_{gi} value located in the numerator of the quadratic equation directly affects the calculated gas in place, so a realistic exponential approximation must be chosen.

Solution Procedure. Rearrange the quadratic equation to the following form.

$$G_a = \frac{1}{2} \frac{\left[2n \pm 2(n^2 + nq_g - nq_{gi})^{0.5}\right]G_P}{q_{gi} - q_g}. \quad\quad (3.10)$$

Dividing into components,

$$n = \frac{1}{2} \frac{q_{gi}}{\left[1 - \left(\dfrac{p_{wf}/z_{wf}}{p_i/z_i}\right)^2\right]} \quad \text{...............} \quad (3.11)$$

$$A = 2(n^2 + nq_g - nq_{gi})^{0.5}, \; MMscf/d. \quad \text{...............} \quad (3.12)$$

Eq. 3.10 can now be shortened to

$$G_a = \frac{1}{2}\left(\frac{[(2n \pm A)]G_p}{q_{gi} - q_g}\right) \quad \text{...............} \quad (3.13)$$

Note, that there are two solutions to the quadratic equation. Select the positive G root given that the negative root is infeasible.

Sum calculated (G_a) with cumulative production (G_p) for each data point and plot the resulting (G) as a function of time. Analyze the plot to arrive at an average OGIP value over the boundary-dominated flow segment (BDF) segment. It is suggested to plot log q vs. log t to identify the BDF segment.

Extrapolating backward results in OGIP at initial time:

$$G = G_a + G_p. \quad \text{...............} \quad (3.14)$$

An example of the Bollycotton Field from Chu et al. (2001) is shown in **Fig. 3.9**. Note the presence of two curves representing calculations with Eq. 3.10 (G_a) and **Eq. 13** (G). The curves are readily divided into transient and boundary-dominated conditions.

Analysis Procedure.

For the analysis procedure do the following:

1. Obtain gas production rate (q_g) and flowing tubinghead pressure (p_{tf}) history. The well must not be producing significant water.
2. Estimate initial pressure (p_i) and gas gravity (γ_g) and fluid properties.
3. Plot (log rate vs. time) and draw a straight line through the data points to estimate q_i. The method is not applicable if a realistic straight line is not evident.
4. Apply a vertical pressure profile program to relate flowing tubinghead pressure (p_{tf}) to flowing bottomhole pressure (p_{wf}).
5. Apply Eq. 3.10 and Eq. 3.11 to calculate OGIP.

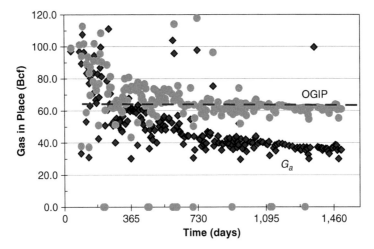

Fig. 3.9—Apparent and original gas in place. Note the ability to differentiate transient from boundary dominated conditions from the plot.

6. Plot apparent gas in place vs. cumulative gas production and interpret the results. Be sure to separate transient from steady state conditions.

Example Problem 3.1—Bollycotton Gas Field. The Bollycotton Gas Field was discovered offshore Ireland. **Table 3.2** lists reservoir properties and OGIP as presented in Chu et al. (2001). Apply the quadratic concept to analyze production history and determine OGIP.

Example Calculations. By inspection of the ln rate vs. time plot, q_{gi} = 54 MMscf/D rate. p_{wf} values are interpreted from the flowing tubinghead pressure (p_{tf}).

Eq. 3.9 has the form $q_g = q_{gi} - \dfrac{2q_{gi}}{\left[1 - \dfrac{p_{wf}/z_{wf}}{p_i/z_i}\right]G} G_P + \dfrac{q_{gi}}{\left[1 - \left(\dfrac{p_{wf}/z_{wf}}{p_i/z_i}\right)^2\right]G^2} G_P^2$.

Reduce the initial condition (p/z) relationship: $\left(\dfrac{p_{wf}/z_{wf}}{p_i/z_i}\right) = \left(\dfrac{p_{wf}/z_{wf}}{1,200/0.91}\right) = \left(\dfrac{p_{wf}/z_{wf}}{1,319}\right)$

Reduce (n) to a single variable: $n = \dfrac{q_{gi}}{\left[1 - \left(\dfrac{p_{wf}/z_{wf}}{p_i/z_i}\right)^2\right]} = \dfrac{54.0 \text{ MMscf/D}}{\left[1 - \left(\dfrac{p_{wf}/z_{wf}}{p_i/z_i}\right)^2\right]}$, in MMscf/D.

Let p_{wf} = 366.7 psi q_{gi} = 54.0 MMscf/D, and q_g = 55.0 MMscf/D.

$$n = \dfrac{q_{gi}}{\left[1 - \left(\dfrac{p_{wf}/z_{wf}}{p_i/z_i}\right)^2\right]} = \dfrac{54.0}{\left[1 - \left(\dfrac{p_{wf}/z_{wf}}{1,319}\right)^2\right]} = \dfrac{54.0}{\left[1 - \left(\dfrac{366.0}{1,319}\right)^2\right]} = 58.5 \dfrac{\text{MMscf}}{\text{D}}.$$

Shorten the general equation to: $G_a = \dfrac{1}{2}\left(\dfrac{[2n \pm A]G_P}{q_{gi} - q_g}\right)$.

Example calculation with p_{wf} = 366.7 psi, when q_{gi} = 54.0 MMscf/D and q_g = 55.0 MMscf/D.

$$A = 2(n^2 + nq_g - nq_{gi})^{0.5} = 2\left\{\left[(58.5^2 + (58.5)(55.0) - (58.5)(54.0)\right)^{0.5}\right]\right\} = 118.0 \dfrac{\text{MMscf}}{\text{D}}.$$

Calculate original gas in place (G).

$$G(+) = \dfrac{1}{2}\dfrac{\left[2n \pm 2(n^2 + nq_g - nq_{gi})^{0.5}\right]G_p}{q_{gi} - q_g} = \dfrac{(2n+A)G_p}{2(q_{gi} - q_g)} = \dfrac{[2(58.5) + 118.0]607.8}{2(54.0 - 55.0)} = 71,417 \text{ MMscf}.$$

$$G(-) = \dfrac{1}{2}\dfrac{\left[2n \pm 2(n^2 + nq_g - nq_{gi})^{0.5}\right]G_p}{q_{gi} - q_g} = \dfrac{[2n - A]G_p}{2(q_{gi} - q_g)} = \dfrac{[2(58.5) - 118.0]607.8}{2(65.0 - 55.0)} = 0.3 \text{ MMscf}.$$

The results of these calculations are plotted for approximate and total gas in place. A good horizontal line is obtained after 700 days of production. *OGIP* = 62 Bscf. The ($G-$) values shown in the last column of **Table 3.3** are obviously incorrect.

p_i = 1,200 psi	γ_g = 0.554	h = 76 ft	G_p = 25.3 Bscf	OGIP = 53.2 Bscf
	T_{res} = 580 °R	ϕ = 22.3%	k = 100 to 109 md	μ_g = 0.013 cp

Table 3.2—Reservoir properties and OGIP listed in the Chu et al (2001) paper.

Time	Rate	p_{wf}	p_{pr}	z	G_p/int	ΣG_p	Rate	p/z	n	A	G+	G−
days	Mscf/D	psia	psia		MMscf	MMscf	MMscf/D	psia		MMcf/D	Bscf	Bscf
365	56315	1010.2	1.48	0.92	204.1	204.1	56.3	1098.0	175.9	354.1	−31.1	0.1
730	55492	999.9	1.46	0.92	202.7	406.8	55.5	1086.8	168.2	337.8	−91.9	0.2
1095	55035	336.7	0.49	0.92	201.0	607.8	55.0	366.0	58.5	118.0	−69.0	0.3

Table 3.3—Partial list of results of calculations.

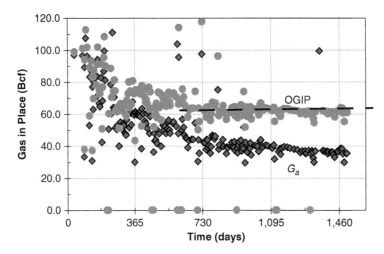

Fig. 3.10—G_a and OGIP plot—example. Measured bottom-hole flowing pressure (p_{wf}) replaced the need for static bottomhole pressures (p_{res}) in the material balance equation.

Multiplot Analysis. Apply the Arps and quadratic concepts to analyze production history.

1. Semilog rate vs. time: review in the context of the Arps curves.
2. Logarithmic rate vs. time: determine transient and boundary dominated segment and the time frame over which they predominate.
3. Rate vs. cumulative production: for this case, because an exponential decline was identified.
4. Find quadratic solution for this particular case because bottomhole flowing pressures were included in the data file and it was a gas reservoir.

Analysis. A reasonably good straight line can be drawn through the 300- to 1,400- day period, which seems to verify the exponential nature of the decline.
The Bollycotton Ga Field declined at 64.5%/yr over the majority of its life.
The straight line approximation intersects the y axis at $q_i = 54$ MMscf/D.

$$D_i = \frac{\ln(54/4.5)}{1,400\,days} = \frac{\ln(54/4.5)}{3.84\,yr} = 64.5\% / \text{yr}.$$

$$EUR = \frac{(54\,MMscf/D)(365\,days/yr)}{0.645\,frac/yr} = 30.6\,Bscf.$$

When comparing apparent gas in place (G_a) and the OGIP (G), note the horizontal nature after approximately 18 months of production for the OGIP (G) estimate while the G_a value continuously declines. OGIP = 62 Bcf.

Summary of Analysis.
Fig. 3.11—Performance: The field appears pressure depleting with a well-defined exponential decline.
Fig. 3.12—Semilog rate vs. time: $D = 64.5\%/y$, EUR = 30.6 Bscf – exponential decline.
Fig. 3.13—Logarithmic rate vs. time plot: Boundary-dominated conditions predominate. Extreme curvature is probably caused by dramatically increasing compressibility for the pressure-depleting case.
Fig. 3.14—Cartesian rate vs. cumulative production: Ultimate recovery = 26.5 Bcf.
Fig. 3.15—Quadratic solution: OGIP = 62 Bcf; ultimate recovery $\approx$ 50%.

Smoothing Variable Production 41

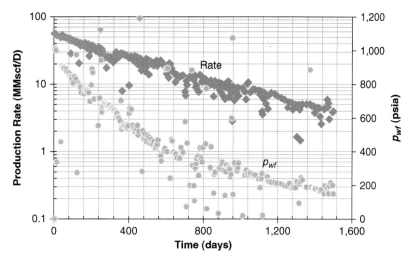

Fig. 3.11—Performance history—Comparing bottomhole flowing pressure and rate decline histories. The apparent lack of significant water production and considerable pressure loss seem to indicate pressure-depletion conditions.

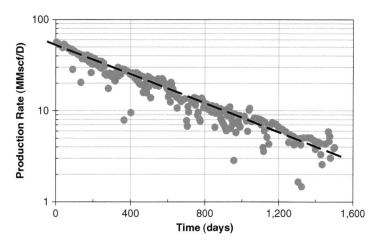

Fig. 3.12—The semilog rate vs. time history seems to follow an exponential decline. What about at early time?

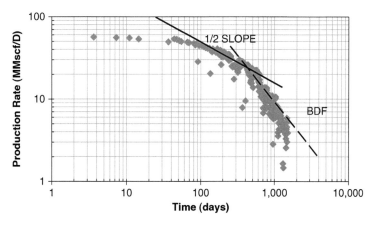

Fig. 3.13—A good transient extrapolation is seen to start approximately 100 days. A logarithmic rate vs. time plot showing well history trending from transient to boundary-dominated depletion conditions near 350 days.

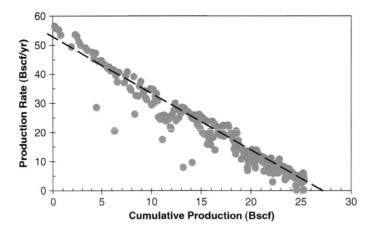

Fig. 3.14—Rate vs. cumulative recovery plot. Note, a good straight line can be constructed over the majority of the performance history after approximately 13 Bscf of gas was produced. EUR = 27 Bscf. Extrapolation to q_i reinforces the semilog rate vs. time plot (q_i = 54 MMcf/D).

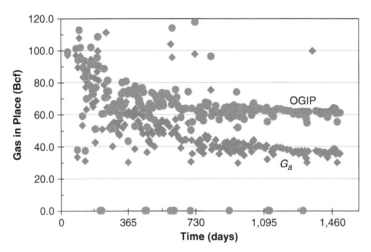

Fig. 3.15—Solution to quadratic equation.

Reservoir Rock and Fluids Properties			
h = 70 ft	p_i = 4,175 psia	γ_g = 0.57	c_t = 1.824x10^{-4} psi^{-1}
ϕ = 6 %	p_{wf} = C = 710 psia	z_i = 0.97	μ_g = 0.0255 cp
*k = 0.07 md	T_{res} = 160°F	z_{pwf} = 0.94	p_{wD} = 0.1702
S_{wi} = 35%	r_w = 0.354 ft	B_{gi} = 0.0041 cf/scf	s = –5.17

Table 3.4—Producing and Reservoir Properties.

Example Problem 3.2—West Virginia Well A Analysis Problem. This gas well was completed in the Onondaga Chert and hydraulically fractured. Adapted from Buba (2003).

Learning Objectives. The learning objectives are

- Realize applying different types of decline curve plots to the solution procedure can really help when interpreting performance.
- Understand the setup, solution, and analysis procedure for applying the quadratic equation.

Analyze the following figures in order to answer the questions that follow.

Fig. 3.16—Logarithmic rate vs. log time. **Fig. 3.17**—Log rate vs. time. **Fig. 3.18**—Cartesian rate vs. cumulative production. **Fig. 3.19**—Quadratic equation, apparent and original gas in place.

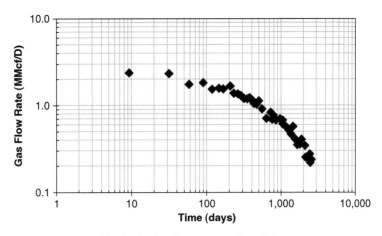

Fig. 3.16—Log-log rate vs. time plot.

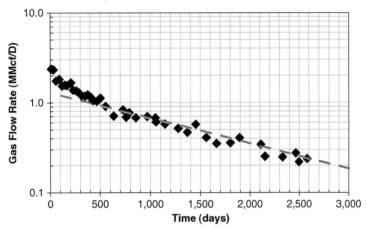

Fig. 3.17—Semilog rate vs. time plot.

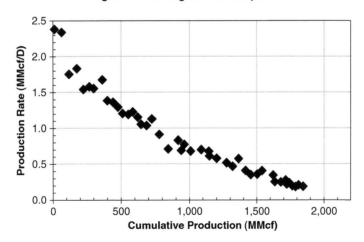

Fig. 3.18—Flow Rate – cumulative recovery plot.

Tasks and Questions.

- Estimate the probable drive mechanism. Is it pressure depletion or waterdrive?
- Divide the producing history into transient and boundary-dominated sections.
- Use the Arps type curve match: $D =$.
- Determine the EUR from the Arps ln the rate vs. time plot.
- Boundary-dominated conditions start at what point?
- Determine q_i and D_i from the exponential curve.
- Estimate the apparent gas in place (G_a) as a function of history.
- Estimate the OGIP (G).

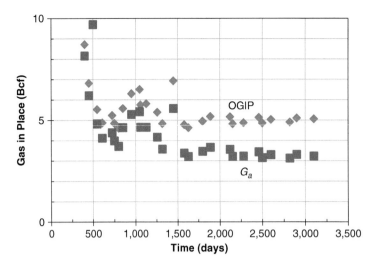

Fig. 3.19—Comparing G_a to sum of G_p plus (Ga – OGIP).

A good quality straight line can be fitted to the data beginning at 1,800 days. Do you agree with this statement? Can you find a straight line in Fig. 3.18?

West Virginia Well A: Summary of Analysis. A summary analysis for West Virginia Well A is presented below.

- Appears to be a pressure-
- depleting gas reservoir with a well-defined exponential decline beginning sometime in the 800- to 1,000-day interval.
- Boundary dominated flow.
- Arps: D = 24.4 %/yr, EUR = 1.87 Bscf or 2.09 Bscf for rate vs. cumulative production plot.
- Boundary dominated conditions start at ≈ 650 days.
- G_p = 1.8 Bscf, OGIP = 5.0 Bscf, Recovery Efficiency = 36%.

Fig. 3.20 indicates the presence of two straight lines, with the break occurring during the 600-to 800-day interval. Which is transient and which is boundary dominated?

$$q_i = 1.42 \text{ MMscf/D}, \quad D_i = \frac{\ln(1.25/0.64)}{2.74 \, yr} = 24.4\% / \text{yr}.$$

$$\text{EUR} = \frac{(1.25 \, \text{MMscf}/\text{D})(365 \, \text{days}/\text{yr})}{0.244 \, \text{fraction}/yr} = 1.870 \, \text{Bscf}$$

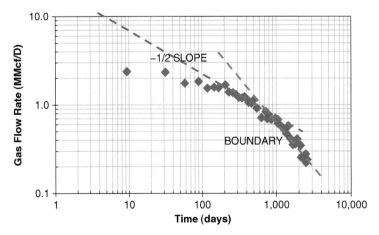

Fig. 3.20—Logarithmic rate - Logarithmic time plot.

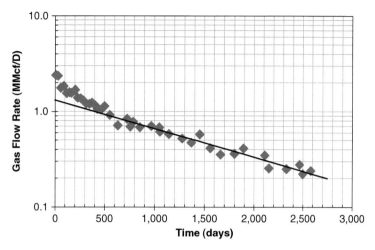

Fig. 3.21—Log flow rate vs. time plot.

The rate vs. cumulative production plot in **Fig. 3.22** determines EUR = 2.09 Bcf.

Is this a qualitative or quantitative estimate?

Fig. 3.23 compares calculated gas in place (G_a) to the sum of the production (G_p) and (G_a), which is equivalent to the OGIP. A straight line begins at 1,350 days indicating OGIP = 5.0 Bcf after producing 1.8 Bcf of gas.

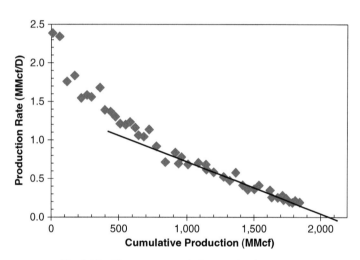

Fig. 3.22—Flow rate cumulative production plot.

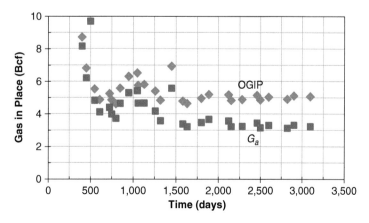

Fig. 3.23—Quadratic plot. OGIP = 5 Bcf.

Chapter 4

Well and Reservoir Models

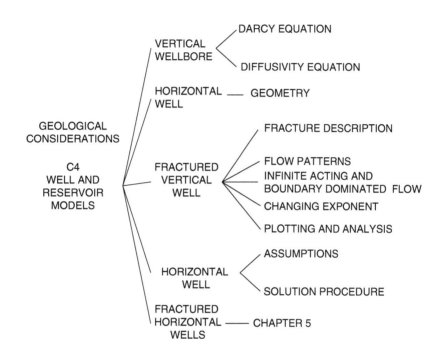

Various reservoir flow conditions may be encountered during oil and gas production operations. Different types of reservoir geologic factors and well completion configurations might result in differing flow patterns in the reservoir, which in turn influence a well's production behavior. Some knowledge of the reservoir geology and well completion configuration is required when interpreting the production performance of a well.

For instance, flow through a well completion is affected by the following factors:

- Producing interval transmissibility. The mobility-thickness product (kh/μ) defines the ability of fluids to move through porous media. Other well completion techniques than vertical well perforations must be considered if the diffusion constant is insufficient to promote commercial production operations.
- Well stimulation techniques, such as hydraulic fracturing or acidizing, and deviated- or horizontal-well drilling are commonly used to increase reservoir flow when reservoir diffusivity is insufficient to sustain commercial production. These completion techniques result in more or improved matrix drainage area open to production.
- Increasing oil rate for viscous-oil wells by multiplying the number of passages between the rock matrix and producing string. Extending the number of potential openings by drilling a horizontal well would materially increase production rate.

In this chapter, the various reservoir and well flow relationships (models) are presented for conventional-reservoir and -well completions. These include well and reservoir models for unfractured vertical wells, vertical wells stimulated by fracturing and unfractured horizontal wells. Chapter 5 presents a more detailed discussion of the well completion techniques and well performance models applicable for unconventional-reservoir completions, often commonly completed with multiply fractured horizontal wellbores.

Geological Considerations

Drilling and completion costs are a capital investment initially incurred to drill the well. Production income must exceed these expenses plus production costs of continued operations for the well to become profitable. This means that the cost of production must be less than the net income generated by production. The well is shut in when production net income equals the production expenses.

Good quality sands with adequate permeability are usually developed with vertical wells because of the ease of drilling and minimization of drilling, completion, and operational costs. Perforated completions in these instances are adequate to permit commercial oil or gas production. On the other hand, vertical fracture-stimulated wells and horizontal-well completions present an expanded drainage area between the matrix face and the wellbore that can materially increase the production when the intrinsic formation permeability is not adequate. Drilling and production costs are generally materially greater with vertical fractured wells and with horizontal wellbores. This is because of the much greater well stimulation costs to create artificial hydraulic fractures or to extend the horizontal drainage reach.

The diffusivity equation is often applied to describe fluid flow in the reservoir under specific flow conditions. Three different drilling and completion arrangements commonly encountered in conventional reservoirs are considered in the following discussion. The application of the diffusivity equation to describe the fluid flow in the reservoir for a horizontal drain hole intersecting multiple fractures (most common in unconventional reservoirs) is addressed in detail in Chapter 5. Each of these flow regimes is exhibited in the transient period of a specific well type. Early-time transient behavior of a very-low-matrix-permeability well, can be distorted by wellbore storage effects, or by reservoir limits and external boundary conditions at late time. In general, we consider

- Applying radial flow equations for the vertical wellbore case.
- Vertical wellbore with hydraulic fractures—Bilinear, linear, and pseudo-radial flow regimes can be applied to the study of fluid flow traveling from the rock matrix face to the fracture, which then conducts these fluids to the wellbore.
- The presence and duration of bilinear or linear flow in the transient behavior of a vertical fractured well is dependent on the dimensionless conductivity of the fracture, which is the ratio of fracture flow to matrix flow. Reservoir bilinear or linear flow transient behavior later transitions to pseudoradial flow behavior provided that the rock matrix permeability is sufficiently high to promote an expanding drainage pattern. Note that pseudoradial flow transient behavior can be exhibited by all fractured wells regardless of the dimensionless fracture conductivity value, provided that the reservoir drainage area is large enough and able to materially expand.
- Un-fractured horizontal wellbore—Initially infinite-acting radial flow to a horizontal wellbore occurs before the onset of reservoir limit effects. The dominant fluid flow pattern in the reservoir is normal to the major axis of the horizontal wellbore. However, it eventually expands to pseudoradial flow like the case of a single vertical fracture model. A pseudoradial flow regime can be exhibited in the well's transient performance if the reservoir drainage area is sufficiently large compared to the length of the horizontal drain hole in the system. Formation linear and pseudoradial flow transient behavior resulting from the production of a horizontal drain hole is occasionally approximated as an ellipsoidal reservoir drainage pattern.

Unfractured Vertical Wellbore Models

Vertical wells are the preferred form of conduit from the oil or gas reservoir to the surface because of the moderate cost and ease of the drilling process. Formation permeability must be sufficiently high to affect the reservoir drainage area in an economic manner. A vertical well may be hydraulically fractured or acidized to increase the production or injection rates when reservoir permeability is too low.

Flow in a cylindrical reservoir toward a perforated vertical wellbore of radius (r_w) occurs in roughly a radial manner, usually defined by the configuration of the outer drainage boundary radius (r_e). Refer to Fig. 4.1.

The diffusivity equation describing radial flow in a circular reservoir indicates that the pressure drop over the drainage radius of the reservoir to the wellbore (from r_e to r_w) occurs in a logarithmic, manner with most of the pressure loss occurring very close to the wellbore.

The Arps (1945) exponential decline ($b=0$) describes fluid flow when the pressure transient has moved through the reservoir and has reached all the boundaries and the reservoir drainage behavior is a function of the reservoir volume. The decline behavior corresponds

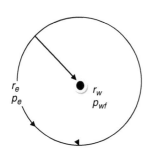

Fig. 4.1—Radial inflow toward the wellbore.

to the late-time transient conditions of fully developed boundary-dominated flow, in which the flow rate is an exponential function of time.

Flow Equations.
Darcy Equation. The analytic expression describing boundary-dominated flow behavior of a vertical well, completed in a closed cylindrical reservoir, is referred to as the Darcy equation for radial flow. Note that the logarithm of the radius term in this relationship is in the denominator of the expression. The other reservoir parameters appearing in this expression are usually treated as constants and form the basis of productivity- index (J) theory or inflow-performance-relationship (*IPR*) calculations. The Darcy equation takes the form

$$q = \frac{0.00708 kh \left(p_r - p_{wf} \right)}{\mu B \left[\ln \left(\frac{r_e}{r_w} \right) - 0.75 \right]} \quad \quad (4.1)$$

The outer boundary of the reservoir drainage area might take any of a variety of shapes, but the pattern is usually assumed to revert to a circular drainage pattern when approaching the perforations. Dietz (1965) shape factors (C_A) are available for a variety of drainage area shapes, well types, and well locations within the reservoir drainage area to account for nonradial flow conditions.

Diffusivity Equation. The diffusivity equation combines the law of conservation of mass, an equation of state describing fluid phase behavior, and a fluid flow law such as Darcy's law to relate the ability of porous media to conduct fluid flow caused by a corresponding pressure loss. A solution to the diffusivity equation is defined by specific initial and inner and outer boundary values.

- Solution is defined by specific limiting factors or boundary condition values.
- Outer boundary can be classified as finite or closed, finite with a partially sealing boundary, finite and constant pressure, or infinite acting.
- Either constant bottomhole-flowing-pressure or constant flow-rate conditions are typically considered at the inner boundary.
- Type of flow condition and completion.

The van Everdingen and Hurst Solutions. van Everdingen and Hurst (1949) developed solutions of the diffusivity equation for radial flow with both constant flow-rate and constant flowing-wellbore-pressure inner boundary conditions. The solutions developed in that work were expressed in terms of dimensionless variables that were related to dimensional field conditions. The dimensionless wellbore pressure and well flow rate cumulative production variables defined by van Everdingen and Hurst (1949) for the constant terminal flow rate and wellbore pressure inner boundary conditions are defined as follows.

Constant rate solution; $pwf(t)$:

$$p_{wD} = \frac{kh \left(p_i - p_{wf} \right)}{141.2 q \mu B} \quad \quad (4.2)$$

Constant pressure solution; $q(t)$:

$$q_{wD} = \frac{141.2 q \mu B}{kh \left(p_i - p_{wf} \right)} \quad \quad (4.3)$$

$$Q_{pD} = \frac{0.8936 N_p B}{\phi c_t h r_w^2 \left(p_i - p_{wf} \right)} \quad \quad (4.4)$$

Dimensionless correlation terms were used in the radial flow analyses to represent the relative reservoir size and elapsed transient time for a particular field condition. The definition of the dimensionless time is expressed in terms of dimensional parameters as follows.

Dimensionless time (field units):

$$t_D = \frac{0.00633 kt}{\phi \mu c_t r_w^2} \quad \quad (4.5)$$

The dimensionless drainage radius is defined as the ratio of the radius of the outermost reservoir boundary (r_e) to the radius of the innermost boundary, [i.e., wellbore radius (r_w)]. The definition of the dimensionless drainage radius therefore furnishes an estimate of the relative drainage volume for a particular reservoir.

Dimensionless drainage radius:

$$r_{eD} = \frac{r_e}{r_w} \quad \quad \quad \quad \quad \quad \quad \quad \quad \quad \quad \quad \quad \quad \quad \quad \quad \quad \quad (4.6)$$

The three major depletion conditions that can occur in a reservoir for a constant wellbore pressure condition at the wellbore are illustrated in **Fig. 4.2.** Dimensionless cumulative production of a vertical well located in a closed circular reservoir with dimensionless drainage radii ranging from 2 to 8 is shown.

1. The smoothly upward trending curve in Fig. 4.2 denotes the presence of infinite-acting transient conditions as the drainage area constantly expands after the well is opened to production.
2. Curved lines define a transition segment caused by the increasing influence of the outer boundary on the reservoir drainage volume. Larger drainage volumes would initiate transition flow at later times along the infinite acting portion of the curve.
3. An essentially horizontal curve after the transition regime occurs when the rate-transient behavior of the well is under the presence of a closed finite reservoir, for which boundary-dominated flow conditions exist.

Constant Rate Solution (p_D). The relationship describing pseudosteady-state pressure drawdown resulting from the constant-rate production of a closed, circular reservoir has been expressed as the equation of a straight line by Ramey and Cobb (1971) whose solution is shown in **Fig. 4.3.**

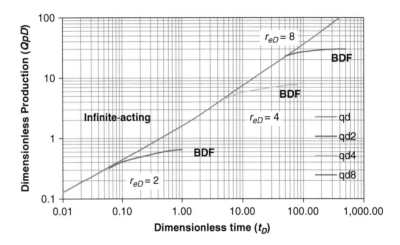

Fig. 4.2—van Everdingen and Hurst (1949) constant pressure solution. The essentially flat lying curve when the production rate is governed by the effects of a closed finite reservoir. Flow conditions can be divided into infinite acting, transition, and boundary-dominated-flow (BDF) cases.

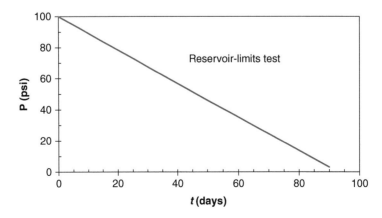

Fig. 4.3—The well-known reservoir-limits test.

$$\frac{kh(p_i - p_{wf})}{141.2q\mu B} = \frac{2}{(r_e/r_w)^2}\left(\frac{0.00633kt}{\phi\mu c_t r_w^2}\right) + \left[\ln\left(\frac{r_e}{r_w}\right) - 0.75 + S\right] \quad\quad\quad (4.7)$$

Each of the terms in the Ramey and Cobb (1971) solution for constant flow rate production may be treated as constants except for p_{wf} and time. Recall the well-known transient-pressure reservoir-limits or drawdown test illustrated in Fig. 4.3.

Constant Pressure Solution (q_D). Fetkovich (1980) adapted the van Everdingen and Hurst (1949) constant pressure solution for a closed, circular reservoir in the form of an exponential equation for production decline curve analyses.

Taking the logarithm of both sides and expressing the solution in the form of an equation of a line result in the following relationship. The constant terminal pressure solution is illustrated graphically in **Fig. 4.4**. Have we seen this semilog rate vs. time plot before?

$$\ln q = \ln\left\{\frac{kh(p_i - p_{wf})}{141.2q\mu B\left[\ln\left(\frac{r_e}{r_w}\right) - 0.75 + S\right]}\right\} - \left(\frac{0.00633kt}{\phi\mu c_t r_w^2}\right)\frac{2}{\left(\frac{r_e}{r_w}\right)^2\left[\ln\left(\frac{r_e}{r_w}\right) - 0.75 + S\right]} \quad\quad (4.8)$$

Horizontal-Unfractured-Well Case

Vertical wells are the preferred completion method when economically feasible because of the decreased cost and complexity compared with deviated or horizontal wells. However, horizontal well completions have often been proved to be a viable option when reservoir permeability is too low and/or the reservoir fluid viscosity is too high to afford economic production with a vertical well completion.

Initially, inflow from the reservoir to the horizontal wellbore will be infinite-acting radial flow toward the horizontal drain hole until the effects of the upper and lower formation bed boundaries are exhibited in well performance. Reservoir flow then transitions into linear flow normal to the principal axis of the horizontal wellbore. Transition from initial infinite-acting radial flow to linear flow and the linear flow behavior of a horizontal well at early transient time tends to appear as generally elliptical in nature. A second infinite-acting pseudoradial flow regime might be exhibited in the transient performance if the reservoir drainage area is sufficiently large compared to the horizontal drain hole length as production time increases. This flow regime is generally analogous to the pseudoradial flow regime that can be exhibited by vertically fractured wells at intermediate transient time. The presence of infinite-acting pseudo radial flow behavior in the transient performance of a horizontal well, might also be considered as an indication that the reservoir permeability is sufficiently high to promote growth of the reservoir drainage area to a more radial or pseudoradial outline.

Extended-reach deviated and horizontal wells are drilled to

- Reduce the surface footprint in the offshore (or a restricted-space) environment. Normally the wellbore may be deviated to normal or near normal when drilling through the target zone.

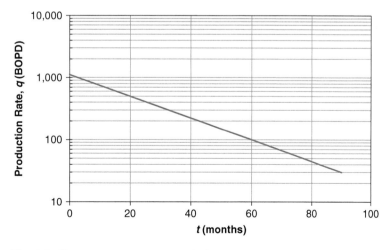

Fig. 4.4—The constant pressure solution. The constant pressure solution results in a (log q vs. t) semilog linear plot that is equivalent to the Arps exponential decline curve.

- Increase the surface area open to production for low permeability formations, provided that the completions are made with a slotted or perforated liner.
- Increase production rates of high-viscosity crudes because the inflow rate per unit cross-sectional area is materially less than that for the vertical well case. Reduced sand and/or water production rates are often also the result because near-wellbore pressure drop has been materially reduced with horizontal well completions.

Horizontal wells in Lake Maracaibo, Venezuela, typically produce at 3- to 4.5-fold greater flow rates than vertical wells producing from the same horizon (Guimeráns et al). Zerpa (1995) found that production rates increased at least threefold with reduced water cuts for horizontal well completions when compared to the case of vertical wells. The heavy oil deposit in that study was reported to have oil gravity of 8° API, with a fluid viscosity of 1,900 cp. The optimal horizontal lateral lengths in that study were found to be between 800 and 1,200 ft.

Dou et al. (2007) and others have applied Arps (1945) and other well performance analysis equations of lines to compare production predictions for both vertical and horizontal wells producing high-viscosity crude oils. Predictions for both well types were found to be of comparable quality.

Horizontal-Well Geometry. The reservoir drainage pattern exhibited for a horizontal well completion at early-to-intermediate transient times is often visualized as essentially being that of an ellipse, because of the nature of the horizontal completion. The early-time elliptical reservoir drainage area will gradually transition into a more radial shape with increasing history. A similar elliptical-to-radial transition effect might also be observed for a higher permeability formation drilled in a high-viscosity-oil reservoir. The high-permeability/high-viscosity case has an effect on the reservoir drainage pattern similar to that of a multiply fractured horizontal well development of a very low permeability shale well completed in an unconventional reservoir. In this case, the rock matrix acts as an impermeable barrier when the mobility $(\lambda=k/\mu)$ of the reservoir region lying outside of the stimulated area is extremely low.

Figure 4.5 presents an idealized view of the outward progression of the reservoir drainage area boundary, from primarily linear flow to the wellbore initially (t_1), to a more pseudoradial condition at t_3. The horizontal wellbore in this idealization is assumed to be drilled at the middle of the formation thickness. Standoff from the bottom of the reservoir pay zone is half of the low-permeability-sand net-pay thickness.

Horizontal Well Example

A field example that illustrates the transient production performance of a horizontal well completion can be found in the results of a study reported by Shih and Blasingame (1995) for a California oil well. The reservoir pay sand is approximately 120 to 150 ft thick, consisting of a highly variable turbidite formation with an effective permeability to oil that reportedly averages less than 3 md, and the oil viscosity is approximately 2 cp. Turbidites are usually highly discordant, with irregular high-permeability streaks that can extend over considerable areas.

Learning Objectives. The learning objectives are to

- Understand how the characteristics of an unfractured horizontal well that has been completed in a permeable formation, but which contains a high-viscosity reservoir fluid, can be interpreted using Arps (1945) methods.
- Realize how field operations can affect performance forecasting. Performance parameters should be determined from the most logical data set.

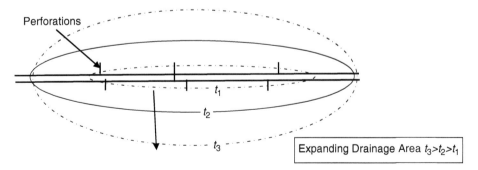

Fig. 4.5—An idealized version of the progression of drainage volume of a horizontal well perforated along the length of the casing. Outward progression of the drainage area eventually culminates in pseudoradial flow if the reservoir is sufficiently large relative to the horizontal wellbore length. ($t_1<t_2<t_3$)

Performance History. Fig. 4.6 divides well performance history into three segments. The three segments are (1) Initial decline and a variable p_{wf}, (2) an orderly decline, and (3) dramatically increasing p_{wf}.

Segment 2 was used to predict performance by assuming well problems could be fixed.

The time vs. cumulative recovery plot in **Fig. 4.7** relates production time to the cumulative oil recovery history. The figure allows the analyst to relate oil production history parameters shown in **Table 4.1.**

The portion of the history attributed to each segment is developed from Fig. 4.6 where a $D = 16.8\%$/yr decline rate was developed by analyzing the semilog rate vs. time plot shown in **Fig. 4.8.**

Conclusions. Application of pseudo production time and pressure drop normalized flow rate analysis techniques can significantly aid in the interpretation of well production performance (**Figs. 4.9 and 4.10**).

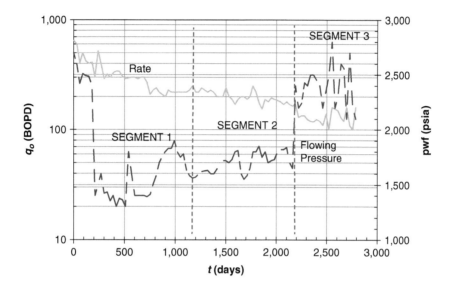

Fig. 4.6—Production history for the subject well. Note the highly variable flowing pressure which seems to control performance. Adapted from Shih and Blasingame (1995).

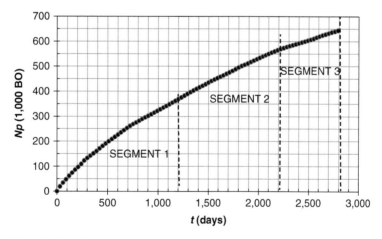

Fig. 4.7—Relating producing time to cumulative production encompassing each segment. Adapted from Shih and Blasingame (1995).

Segment	Time, t (days)	Cumulative Production (1,000 BO)	Rate (BOPD)	Description
	Fig. 4.6	Fig. 4.7	Fig. 4.8	
1	0 → 1200	0 → 365		Transient?
2	1200→2200	365 → 560	250→ 175	Boundary? moderate (p_{wf}) variation
3	2200→2800	560 → 630		Erratic (p_{wf}). Well producing water?

Table 4.1—Segment properties.

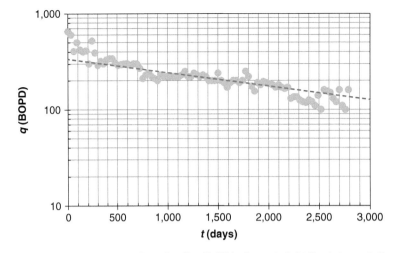

Fig. 4.8—Semilog rate vs. time plot. D = 16.8%/yr for a straight-line interpretation over the 1,200 to 2,200-day segment 2 history. Adapted from Shih and Blasingame (1995).

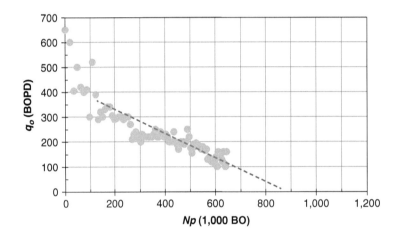

Fig. 4.9—Field data plot and fitting the straight line for extrapolation to obtain the estimated ultimate recovery (EUR). EUR= 870 MBO. Adapted from Shih and Blasingame (1995).

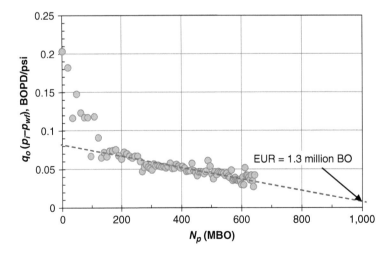

Fig. 4.10—The normalization process smoothed the erratic data in Segment 3. Adapted from Shih and Blasingame (1995).

Fractured Vertical Wellbore Case

Vertical and horizontal wells that are completed in low permeability reservoirs are commonly hydraulically fractured to expose additional reservoir rock surface to yield an increase in the production rate of the well. **Fig. 4.11** is an illustration that depicts the geometrical relationship of a vertically fractured well in a single-layer reservoir.

The production performance of vertical fractures has been investigated in many studies over the past 50 to 60 years. These studies have included investigations to understand the pressure- and rate-transient behavior of uniform-flux, infinite-conductivity, and finite-conductivity vertical fractures. The uniform-flux and infinite-conductivity idealizations for vertical fractures are often used to develop useful analytic solutions for interpreting the transient behavior of vertically fractured wells.

All fractured wells in fact possess a finite conductivity ($k_f b_f$). The fracture conductivity may be very high, but it still has a finite value. There will be a negligible pressure drop in the fracture when the contrast between the fracture conductivity and the product of the reservoir permeability and the fracture half-length (kX_f) is sufficiently high. In such a case, the vertical fracture will behave as though it has essentially an infinite fracture conductivity. As Gringarten et al. (1974) have demonstrated, an infinite-conductivity-vertical-fracture response can be effectively obtained using the uniform-flux-vertical-fracture solution evaluated at a spatial position in the fracture that is 0.732 of the fracture half-length from wellbore toward the fracture tip.

Dimensionless Fracture Conductivity. The dimensionless fracture conductivity (C_{fD}) is one of the key parameters governing the transient behavior of a vertically fractured well. As it is currently defined and used for the pressure- and rate-transient analyses of vertically fractured wells, the dimensionless fracture conductivity was originally reported by Agarwal et al. (1979) in the study of the effect of hydraulic fracturing on production. The dimensionless fracture conductivity value relates the effect of fracture conductivity ($k_f b_f$) to the product of reservoir permeability and fracture half-length (kX_f) product. As can be observed in the definition of the dimensionless fracture conductivity, a low dimensionless conductivity fracture can be obtained with a high permeability formation or a very long fracture.

$$C_{fD} = \frac{k_f b_f}{kX_f} \quad \quad \quad (4.9)$$

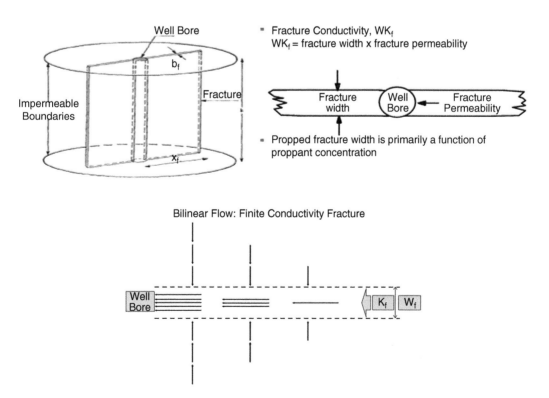

Fig. 4.11—Schematic of geometrical relationship for a vertically fractured well. Adapted from Fekete (2012).

Where the terms in Eq. 4.9 are
b_f = fracture width, ft;
X_f = fracture half length (L_c), ft;
k_f = fracture permeability, md;
and k = reservoir permeability, md.

The definition of dimensionless fracture conductivity given in Eq. 4.9 is related to other forms of the relative fracture and reservoir flow capacity of vertically fractured wells such as the fracture capacity (α) defined by Prats (1961) and relative fracture conductivity (C_r) reported by Cinco-Ley et al. (1978).

$$\alpha = \frac{\pi}{2C_{fD}} \quad\quad\quad (4.10)$$

$$C_r = \frac{C_{fD}}{\pi} \quad\quad\quad (4.11)$$

Bello and Wattenbarger (2008) indicated that a fracture may be treated as essentially infinite conductivity when $C_{fD} > 50$.

Cinco-Ley, et a.l (1978) stated that the pressure transient behavior of vertical fractures is essentially the same for $C_{fD} > 20\pi$.

Various authors have reported threshold values for infinite-conductivity-vertical-fracture transient behavior, with a reasonable value of $C_{fD} \geq 300$ generally considered to be adequate for infinite-conductivity-fracture behavior.

Fig. 4.12 illustrates the geometric and linear relationship between matrix and fracture flow. Adapted from Fekete (2012) and Holditch (2014).

Guppy et.al. (1981) investigated the rate-transient behavior of vertically fractured wells and proposed a specialized analysis procedure for bilinear flow well performance. A graphical analysis of the reciprocal flow rate vs. the quarter root of time ($1/q$ vs. $t^{1/4}$) was demonstrated to have a slope that is proportional to the squared product of formation permeability and fracture conductivity. Note that the lower the dimensionless conductivity of the fracture, the more bilinear flow behavior exhibited in the well's early transient performance.

Similarly, the linear flow transient behavior of a vertically fractured well can be evaluated using a reciprocal flow rate analysis referenced to the square root time ($1/q$ vs. $t^{1/2}$). The slope of the graphical analysis straight line during linear flow of a moderate-to high- dimensionless-conductivity fracture is proportional to the squared product of formation permeability and effective fracture half-length. Higher dimensionless conductivity fractures exhibit more linear flow behavior than more moderate or lower dimensionless conductivity fractures.

Knowledge of the transition from infinite-acting transient to boundary-dominated flow conditions of a vertically fractured well is important because this behavior provides first indications of reservoir drainage area and reserves size. Production decline analyses have been reported in the literature. The analyses have considered only transient fracture flow behavior, infinite-acting transient, and transition flow behavior. The analyses have also

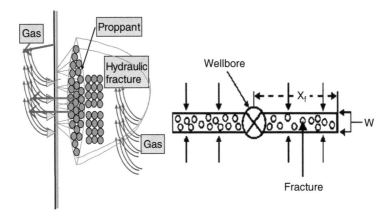

Fig. 4.12—An example reflecting flow from the matrix to the hydraulic and fluid-filled fracture where the fluid is transported to the wellbore by bilinear flow. Fracture permeability is increased by the introduction of a proppant to keep the faces apart when pressure is released after a hydraulic fracture treatment. Natural fractures do not contain proppant material. Adapted from Fekete (2012).

considered studies that encompass all the production performance history of a vertically fractured well, including the transient, transition, and boundary-dominated flow behavior. One such study that investigated the production performance of vertically fractured wells in single- and dual-porosity systems is Aguilera and Ng. (1991). Analytic relationships for the pressure- and rate-transient behavior of a vertically fractured well during the principal flow regimes have been reported by various authors. A general summary of these rate- and pressure-transient relationships for vertically fractured wells can also be found in Azari et al. (1991).

Changing Flow Patterns. A vertical fracture intersecting a well increases and alters the reservoir drainage flow pattern from what would normally be a radial to more linear or modified-linear flow behavior. Initially the early transient behavior of wells completed in finite reservoirs will exhibit infinite-acting transient flow that will eventually transition to boundary-dominated flow at late transient time. In general, production at the wellbore consists of expansion of fluids in the vertical fracture and inflow contribution from the reservoir matrix connected to the well by the vertical fracture.

The contribution of each segment in the system is subject to respective permeability values and is also a function of the time frame over which each occurs and predominates. Cinco Ley et al. (1978) divided the transient behavior resulting from the production of a vertical hydraulically fractured well into five distinguishable flow regimes. **Fig. 4.13** is an illustration that depicts this progression through the principal transient flow regimes that can be exhibited by production through a vertical fracture.

During this time;

- Maximum production rate occurs during the fracture linear flow and bilinear flow as a result of wellbore storage and fracture unloading.
- Formation linear flow production is derived from the formation matrix. The fractures simply act as conduits to the wellbore. This flow regime predominates in the transient performance of moderate- to high- dimensionless- conductivity vertical fractures.
- Pseudoradial flow might occur. Matrix permeability is sufficient for the outer drainage area boundary to progress outward and become more circular in outline. This system is seldom encountered in very-low-permeability reservoirs before production rate reaches the economic limit.

Introducing different flow regimes requires reordering of mathematical equations to properly model a particular part of the flow system. The additive nature and uncertainty of time-value predominance of the different processes increases the complexity of describing reservoir conditions.

A Typical Performance History. Fig. 4.14 illustrates the difficulty often encountered in applying the flow solution to a typical production decline of a hydraulically fractured well completed in a low permeability reservoir. Infinite-acting reservoir conditions occur initially while boundary-dominated (pseudosteady state) flow is exhibited when the drainage volume provides the sole source of production. What happens when both pressure responses occur simultaneously?

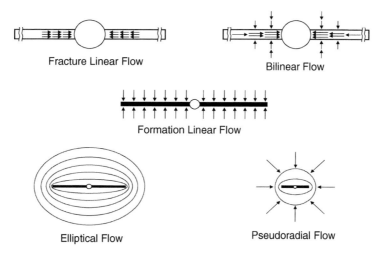

Fig. 4.13—Outline of the flow systems that can occur during the depletion of a fractured well. Adapted from Cinco-Ley et al. (1978).

Changing Values of *b*-Exponent. Conventional decline curve analysis techniques do not necessarily apply for the fractured vertical and horizontal well cases. The single radial flow regime has now become a series of changing flow conditions over the production history of the fractured vertical or horizontal well. For instance, **Fig. 4.15** indicates that a fractured, dual porosity flow system might change in the following bimodal manner.

- Early-time exponential decline ($b = 0$) extrapolations underestimate reserves because the high-permeability-fracture unloading predominates the flow system.

    The extraordinarily low matrix permeability causes rapid production falloff after fracture unloading. Decline dramatically moderates at later time when flow is derived principally from the matrix.

- The nonconverging $b > 1$ exponent results when a dramatic production loss is achieved early in the well life. Production levels off when the rock matrix is the sole supplier of fluid after a steep initial decline. The nature of Arps' equation when $b > 1$ results in a nonconvergence to the abscissa as the Arps equations postulate. Reserves can be overestimated when this happens. Eventually, production should become exponential in nature as boundary-dominated flow is exhibited.

Most interpreters try to fit some equation of a line to model production history. However, both transient and boundary affects contribute in a nonlinear changing manner during the transition middle portion. Remaining reserves are often forecast on the basis of complete line, whereas in actuality predictions should reflect only boundary-dominated flow conditions.

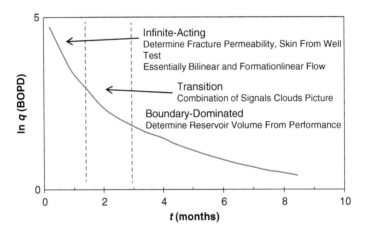

Fig. 4.14—The three major segments of a typical decline curve. There is not an adequate analytical method available for calculating the well flow rate transient behavior that covers the transition period between the transient and boundary dominated flow regimes.

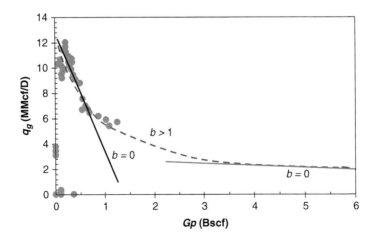

Fig. 4.15—The journey of the Arps (1945) *b*-exponent for a typical hydraulically fractured well. When does the well cease producing if the nearly flat rate remains above the economic limit?

It seems that a better method would attempt to identify the production time frame over which infinite-acting and boundary-dominated flow conditions occur. Then determine the decline characteristics of the system when only the reservoir volume is furnishing the well performance character. Type curve analysis approaches attempt to identify from historical well performance the first indications of boundary effects, improving estimates of future performance.

The following subsection presents a method for applying simplified plotting and analysis techniques to get a good handle on reserves and producing characteristics. The discussion initially concerns determining

- The existence and time frame of the infinite-acting and boundary dominated production segments.
- The flow characteristics from the transient side.
- Reserves from the boundary dominated production side.

Plotting and Analysis Methods. The following discussion concerns the various plotting systems that can be applied to analyze segmented production plots.

Log Flow Rate vs. Log Time Plot. Maley (1985) stated that a log flow rate vs. log time plot results in a straight line for the infinite-acting, linear flow

$$\ln q = -\frac{1}{2}\ln t + \ln C. \quad \quad \quad (4.12)$$

In **Table 4.2**, there are shown the suggested plotting functions.

Fig. 4.16 depicts the relationship between infinite-acting linear flow at early time and boundary-dominated flow at late time on a complete well history.

Divergence from the half slope line denotes the end of completely infinite-acting reservoir behavior and the early onset of boundary effects. Bending of the curve is a function of the contribution of multiple flow systems affecting well performance.

Skin effects, mechanical problems, liquid loading, or ineffective fracture treatments can mask the half-slope behavior during the early transition period. Refer to the superposition discussion in Chapter 3. Curvature of the line during the transition period reflects the difference between fracture and reservoir matrix flow systems. Determining the two straight lines can often be somewhat problematic for variable-permeability producing intervals.

Log Flow Rate vs. Log Time Plot	
"x" plotting function, ln t	Slope of the line, $m = -\dfrac{1}{2}$
"y" plotting function, ln q	"y" intercept = ln "c"

Table 4.2 — Plotting and outcome functions.

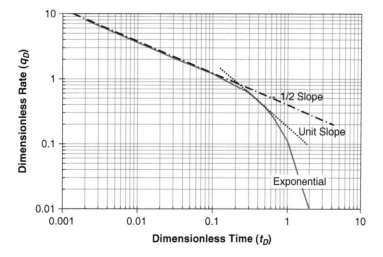

Fig. 4.16—The log-log flow rate vs. time plot can be used to identify the presence/absence of infinite-acting, transition, and boundary dominated flow conditions. The half slope identifies the transient linear flow, while unit slope identifies the boundary dominated flow trend at late production time.

Reciprocal Flow Rate vs. Square Root of Time Plot. Anderson et al. (2010) states that the ($1/q$ vs. $\sqrt{t}$) plot results in a straight line for the infinite-acting linear flow case. The coefficients A and B in the reciprocal flow rate vs. square root of time analysis, Eq. 4.13, are specific functions of the well type (vertical fracture or horizontal well, or multiply-fractured horizontal well), the wellbore pressure drawdown ($p_i - p_{wf}$), and the reservoir and well completion properties. The reservoir properties included in these coefficients are the permeability, net pay thickness, porosity, total system compressibility, and fluid viscosity. The well completion parameters include the system characteristic length (L_c) and steady state skin effect (S).

$$\frac{1}{q} = A\sqrt{t} + B \quad\quad\quad\quad\quad\quad\quad\quad\quad\quad\quad\quad\quad\quad\quad\quad\quad\quad\quad\quad\quad\quad\quad\quad\quad\quad\quad\quad\quad\quad\quad\quad (4.13)$$

Fig. 4.17 provides an example of the application of the square root time plot for the Anderson et al. (2010) well. The change of flow system from transient infinite-acting to boundary-dominated flow occurs at the intersection of the two straight lines indicated in this figure.

Wattenbarger, et al. (1998) demonstrated the half-slope behavior that is often observed on a $1/q$ vs. $\sqrt{t}$. plot signals a change from bilinear to linear flow or from linear to boundary-dominated flow in the flow behavior of the system. Anderson et al. (2010) and Wattenbarger et al. (1998) observed that the same phenomenon occurs for both constant flow rate and constant wellbore flowing pressure cases.

Fig. 4.18 presents the log flow rate vs. log time analysis.

Differentiation between the flow regimes is obvious, with very little transition flow regime exhibited between the infinite-acting linear flow boundary-dominated flow regimes.

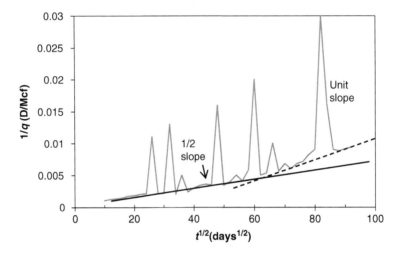

Fig. 4.17—Boundary effects become evident at about $\sqrt{t}$ = 60 or t = 3,600 d. (Anderson, et al. 2010).

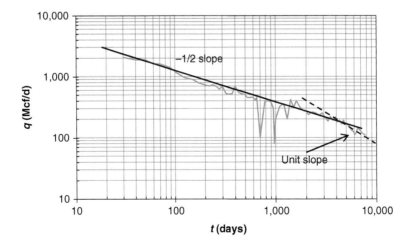

Fig. 4.18—Which line would one use to predict future performance? Data from (Anderson, et al. 2010).

Effect of Formation Damage. Formation damage caused during hydraulic fracture treatments can result in water blockage in the matrix at the fracture/matrix interface. Buildup of water saturation close to the fracture face should be reduced as much as possible to return the rock pore volume and fluid saturations to as close to original conditions as practical. Lower initial flow rates might result if the injected and imbibed water blockage is not removed.

Fig. 4.19 illustrates how significant formation damage flattens the slope of the infinite-acting transient production rate curve. This is done by inhibiting production at early time, when the flow rate should be at a maximum under optimal conditions.

The uppermost curve in Fig. 4.19 defines the expected flow rate behavior when damage is absent. This same distortion was observed in the (t_p/t) ratio plots discussed in Chapter 3.

Fig. 4.20 reproduces the $(1/q$ vs. $\sqrt{t})$ plot for the results of a simulation study conducted by Wattenbarger et al. (1998). Note that theoretically, the straight line should pass through the origin for no inhibiting skin effect. However, in actual practice field cases often intersect above the origin, particularly for large values of damage skin.

Analysis Procedure.

1. Gather all production data (i.e., oil, gas, and water production; flowing bottomhole pressure; and any shut-in pressures).
2. Plot these data as a function of time and interpret the general characteristics and identify the number of possible segments.
3. Construct a semilog flow rate vs. time plot to recognize possible segments. Backward extrapolate to $t = 0$ to determine q_i. Calculate exponential decline rate, if a straight line is justified.
4. Construct a Cartesian flow rate vs. cumulative recovery plot and determine EUR if a reasonable straight line is interpreted.
5. Construct a (log flow rate vs. log time) plot to determine whether transient infinite-acting flow and boundary dominated flow behaviors are present. Recall that ½ slope signifies linear flow behavior while a unit slope signifies boundary influences.
6. Construct a $1/q$ vs. $\sqrt{t}$ plot to determine presence of a ½ slope linear behavior.

Example Problem 1—Vertical Well. Well 19 was drilled horizontally, hydraulically fractured, and completed in a highly variable low permeability rock section located in the Permian Basin. Reserves equaling 44,640 BO were projected from the semilog rate vs. time plot shown in **Fig. 4.21.** Note that the analyst fit a straight line through the 390-day production history. Apply some of the recently discussed principles to determine the reliability of the estimate.

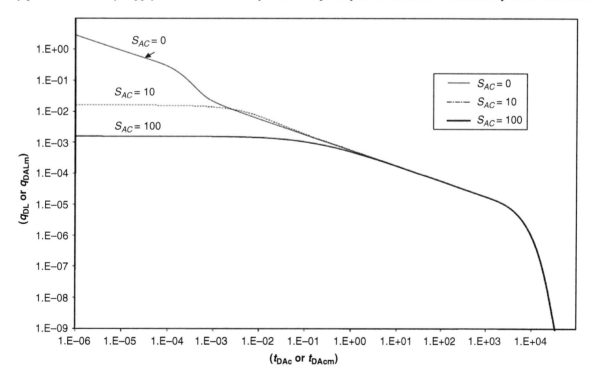

Fig. 4.19—A log flow rate vs. log time plot illustrating effects of formation damage (skin) on the shape of the production rate curve. Expected early time flush production is nullified when skin damage is present. (Wattenbarger et al. 1998).

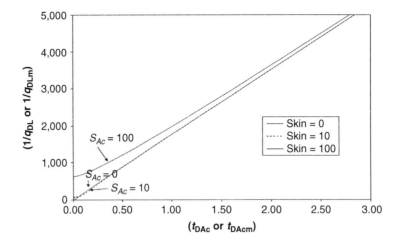

Fig. 4.20—The effect of formation damage (skin) on the shape of the ($1/q$ vs. $\sqrt{t}$) plot. The line passes through the origin when damage skin effect (S) is at a very low value. (Wattenbarger et al. 1998).

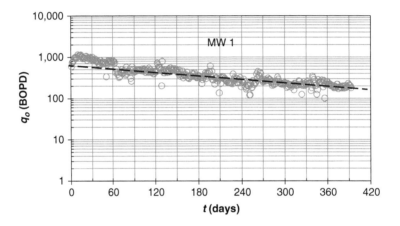

Fig. 4.21—A good straight-line extrapolation of the entire producing history of Well 19. D = 137%/remaining reserves credited to a 5-BOPD economic limit = 44,640 BO (Wattenbarger et al. 1998).

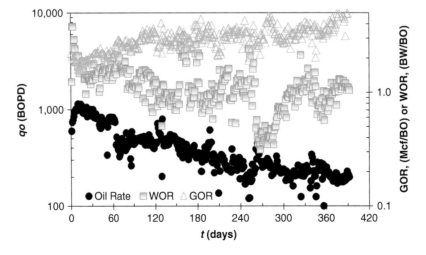

Fig. 4.22—Well 19 production history. Adapted from Wattenbarger et al. (1998).

Learning Objective. The objective of the exercise is to review and interpret multiple plots to determine well character and test the validity of our study.

Fig. 4.22 shows the 390-day production history for Well 19 along with an interpretation of performance. The interpretation performance for Fig. 22 shows that the

- Gas/oil ratio (GOR) rapidly increased to approximately 5,000 to 6,000 scf/STB, which approximates field average in the area. Water production remained moderate and averaged 0.9 BW/BO after approximately 180 days.
- Oil production rate displayed a smooth decline.

The reader's job is to verify the reserves estimate. Do you agree with this study? In your case, apply graphical analyses applicable to study horizontal well behavior to aid in your interpretation. In your case, additional graphical analyses that are applicable for the production analyses of vertically fractured wells should also be applied to aid in your interpretation. Study Figs. 4.23 and 4.24 to aid in your conclusions.

Draw a straight line over the verified boundary-dominated portion of the well history. (**Fig. 4.25.**)

q_i = 600 BOPD, q_o at 390 days = 180 BOPD, N_p at 5 BOPD, N_p at economic limit (EL) = 90,860 BO

Conclusion. Analysis of the full production history revealed that interpretation of early time data alone significantly underestimated the EUR. Additionally, there is an indication at early time that formation damage or an ineffective fracture treatment is likely restricting the initial flow rate which can be observed from **Fig. 4.26.** at early times Therefore, a negative intercept is presented on the $1/q$ vs. $\sqrt{t}$ plot for the boundary-dominated-flow (BDF) region, while early data ($t^{1/2} < 10$) show a positive intercept.

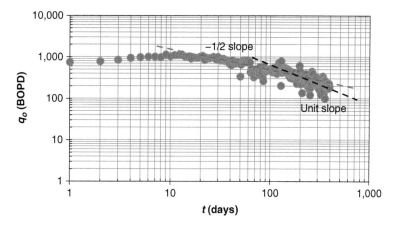

Fig. 4.23—Log flow rate vs. log time plot. Shifting to boundary-dominated flow regime is not obvious. What could this observation be an indicator of?

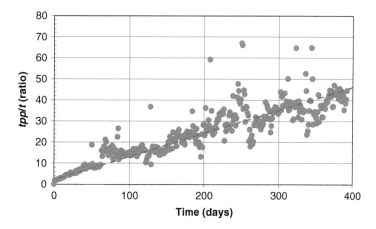

Fig. 4.24—Time ratio plot. More definitive than log rate vs. log time plot. Seems to verify selection of transient time to pseudotime boundary at 14–160 days.

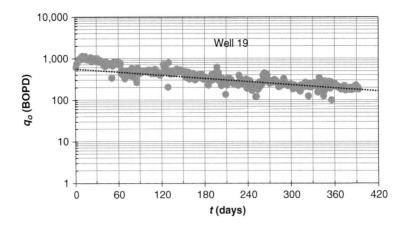

Fig. 4.25—Semilog flow rate vs. time plot. Choosing an exponential fit encompassing the boundary-dominated portion of the production history, $D = 70.3$ %/yr.

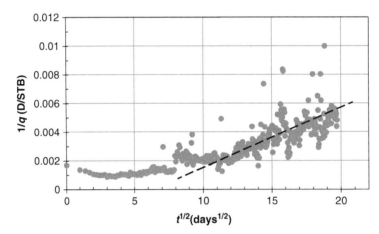

Fig. 4.26—The reciprocal flow rate vs. square root of time linear flow analysis indicates boundary-dominated conditions begin at approximately 140 days.

Problem

Problem 4.1. Compare pressure profiles for linear and radial flow.

Linear Flow: $q = 1.127 L \dfrac{kA(P_{res} - P_{res})}{\mu L}$, $p = p_{wf} + C(L)$.

Radial Flow: $q = \dfrac{0.00708 kh (P_{res} - P_{res})}{\mu B \left[\ln\left(\dfrac{r_e}{r_w}\right) - 0.75 \right]}$ $p = p_{wf} + C \left[\ln\left(\dfrac{r_e}{0.25}\right) - 0.75 \right]$.

Rewrite the Darcy equation for liquid flow to calculate and graph (**Fig. P4.1.1**) the pressure profile for a well draining a 40-acre circular reservoir when $r_e = 745$ ft, $r_w = 0.25$ ft, and $p_{wf} = 1{,}200$ psia.

The constant reservoir parameters are collected and found equal to $C = 306$ psia for the radial flow equation. Solve the equations for a constant (p_{wf}) solution. Assume all reservoir identifiers are equal to C.

$$p = p_{wf} + \dfrac{q\mu B \left[\ln\left(\dfrac{r_e}{r_w}\right) - 0.75 \right]}{0.00708 kh} \quad p_{wf} + 306 \left[\ln\left(\dfrac{r_e}{0.25}\right) - 0.75 \right].$$

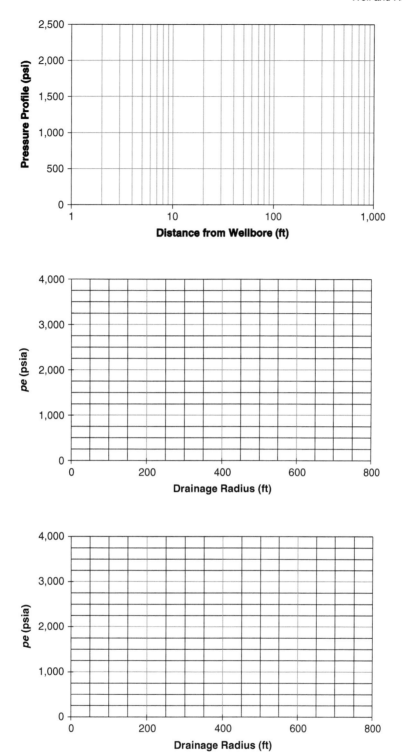

Fig. P4.1.1 — Comparing pressure-drop profiles for linear and radial flow.

Chapter 5
Fractured Horizontal Wells

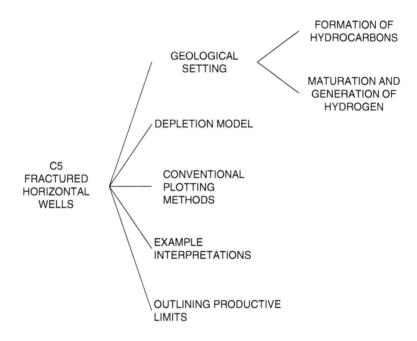

Fractured horizontal well completions are commonly implemented to develop very-low-permeability rocks containing organic-rich shale, siltstones, and fine-grained sand in clastic and carbonate reservoirs. These potentially productive sediments are often identified as "shales" even though they are composed of a variety of rock types. The alternative term "mud-rock" is also used to emphasize the very small grain size. Hydrocarbons are formed in place and must contain the correct combination of kerogen type, maturity level, shale type, source rock average total organic content, geomechanical properties, silica content, sufficient porosity, net pay and permeability to produce oil and gas commercially.

Both the geological setting and the reservoir fluid flow dynamics affect recovery and performance of these multiple-fractured horizontal well completions. The following discussion considers the various aspects for evaluating the potential for these reservoir types.

Geological Setting
Increasing oil and gas prices have afforded development of previously uneconomic very-low-porosity and- permeability sediments. In this case, field development is not a function of structural relief and gravity segregation but depends upon generation of in-situ organic material to form oil and gas deposits. Capillary effects retain these fluids in the very-low-porosity sediments. The environment of deposition to a large extent defines the type of kerogen, that in turn defines the tendency of the organic material to form oil or gas. Productive shales require rapid burial to exclude oxygen.

Oil- and gas-saturated shales are abundantly distributed in the geological section. Berg (1986) discussed how sediment transportation velocity controls the depositional process. Extremely small size particles tend to be

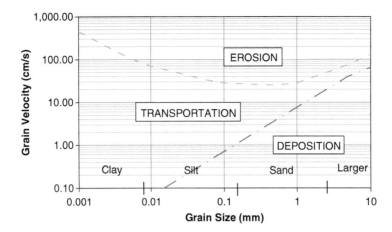

Fig.5.1—Current-velocity and grain-size effects on the erosion, transportation, and deposition envelopes. A minor decline in transport energy can cause deposition of the large-grained material such as coarse sands while fine-grained-material movement might be affected over a wide range of conditions. Modified from Berg (1986).

distributed over a wide variety of depositional environments. **Fig. 5.1** shows how fluid velocity and grain size control the movement of particles, leading to erosion, transportation, and ultimately deposition.

There is little difference in the erosion/deposition boundaries for larger-sized material (greater than coarse-sand size). High current velocities are required to erode greater-than-sand-sized sediments, but small velocity reductions might then cause immediate deposition. Therefore, accumulation usually results in highly porous reservoirs of limited areal extent. On the other hand, **Fig. 5.1.** shows how clay- and silt-sized sediments require high-energy erosional environments. Once mobilized, the sediment might be transported most readily by exceedingly low transport velocities. These sediments commonly occur in lakes, deeper ocean basins, and shelf-area sedimentary environments. Extensive accumulations of very-fine-grained material are usually found in shelf and basin environments where environmental conditions are widespread. Short-term higher energy levels might bring larger-grain-size sediments—i.e., silt and sand—into the depositional site.

Carbonate sediments developed by evaporation and chemical precipitation of seawater are often associated with fine-grained clastic material. These fine-grained sediments are usually rich in organic material and can accumulate over a wide areal extent within these quiescent areas. Permeability in these fine-grained organic materials usually is on the order of from 10^{-6} to 10^{-4} md.

Carbonates are highly susceptible to alteration by movement of ground water since calcium carbonate is readily water-soluble. Erratic solution and dissolution over geologic time can develop "sweet spots" within the reservoir that might be encountered.

Stress relief develops natural fractures, that are common to carbonates. Encountering a high permeability conduit in the low permeability reservoir increases productive matrix face area and fluid storage. Encountering an unforeseen natural fracture often results in unexpectedly high production rates and increased reserves.

Higher than normal production can generally only be established by predicting trends of these "sweet spots." Predicting production performance in these reservoir types presents a particular problem because of the uncertain drainage configuration and the extended transient rate condition occurring after production begins. The following summarizes some of the potential problems with identifying and forecasting production performance.

- Interwell correlation of productive units is extremely difficult because of the erratic nature of the sedimentary interval. Subtle differences in well log definition can signal apparent major changes in rock properties that can greatly affect production performance.
- Unforeseen natural fractures are usually formed with an orientation parallel to strike; they are usually randomly distributed.
- Static pressures require an inordinate amount of shut-in time to attain stabilized conditions in the low permeability sediments. Material balance methods are also generally not effective when transient pressures exist for an extended period.
- Expected reservoir properties must be included to study reservoir conditions with a numerical or analytical simulator. Vagueness of the model reservoir parameters usually precludes a realistic input of the model characteristics.

Classification of the Potential of Various Organic Material Types

Type I—Generally refers to lake deposits containing algae and plankton reworked by bacteria
- Rich in hydrogen (H_2) and poor in oxygen (O_2): H/C > 1.25, O/C < 0.15
- Best material for converting to smaller oil and gas molecules, but not widespread because of limited areal extent at the depositional sites
- Commonly found in oil shales, some of that might be high in sulfur

Type II—Derived from moderately deep marine sediments; usually contains plankton-like material reworked by bacteria
- Rich in hydrogen (H_2) and low in carbon (C): H/C < 1.25, O/C < 0.036
- Tends to generate oil and gas by increasing burial and maturation

Type III—Derived principally from swamp material by oxidation of woody material
- Low in hydrogen (H_2) and high in (O_2): H/C < 1.00, O/C 0.03 to 0.3
- Generates principally dry gas

Type IV—Reworked old deposits
- High in carbon (C) and low in hydrogen (H_2): H/C < 0.5
- Not a widespread category and seldom of interest

Table 5.1—O/C=Oxygen/Carbon Index Ratio. Langmuir (1918) Classifications.

Hydrocarbon Generation. Oil or gas deposits found in very low permeability organic-and hydrogen-rich sediments are derived from post-depositional processes converting organic material into high molecular weight kerogens and into oil and gas accumulations. These sediments must possess organic material containing significant quantities of carbon for the development of kerogen (Hutton et al.1994).

Kerogen type often varies geologically and geographically according to the variety of organic compounds present in the sediments. Bitumen is a kerogen soluble in organic solvents. Insoluble kerogen eventually converts to oil and gas molecules after maturing at elevated pressures and temperatures, usually the product of continued burial and compaction. Eventually, evolved hydrocarbons lodge into rock matrix pore spaces and reside as oil and gas molecules. These molecules are adsorbed on organic material or stored in pore spaces according to their independent saturations. Adsorption is the ability of a molecule to adhere onto an active molecular site. The adsorption quantity is a function of temperature and pressure. Langmuir (1918) proposed an isothermal theorem equating the dynamic equilibrium existing between adsorbed gaseous molecules and free gaseous molecules.

The hydrogen/carbon (H/C) ratio value serves as the preliminary indicator of the hydrogen quantity needed to convert kerogen to a gas and/or liquid hydrocarbon. Conversely, materials with a very high carbon/hydrogen (C/H) ratio are solids, such as coal.

Table 5.1 refers to the following classification of the potential of various organic material types to eventually form into hydrocarbon compounds.

Chemical reactions modify the composition of the material to increasing hydrogen content, that with maturation will eventually convert to graphite as the end product (Seewald 2003).

The modified van Krevelen (1950) diagram shown in **Fig. 5.2** provides a crossplot of (H/C) index atomic ratio on the ordinate axis and (O/C) index ratio on the abscissa as a function of kerogen type envelopes. A glass-like substance (vitrinite) forms the end product of the conversion process. The R_o envelopes measure the amount of vitrinite present in a sample. The figure is used to assess the origin and maturity of oil and gas accumulations. Note how different types of organic material are more likely to reduce to

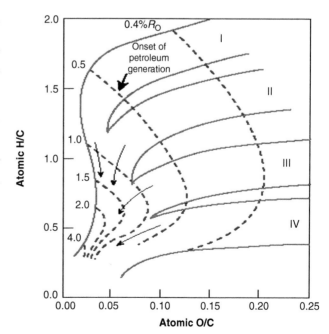

Fig. 5.2—A modified van Krevelen plot reflecting the effect of increasing temperature and pressure on the conversion of kerogen to oil and gas byproducts. Adapted from Boyer et al. (2006).

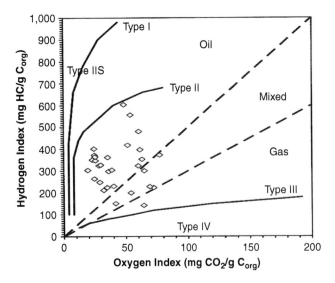

Fig. 5.3—Results of laboratory analysis of rock samples shown as diamonds are plotted on the Van Krevelen diagram. The results seem to indicate the probability of oil-prone production for this case. Adapted from Dembicki (2009).

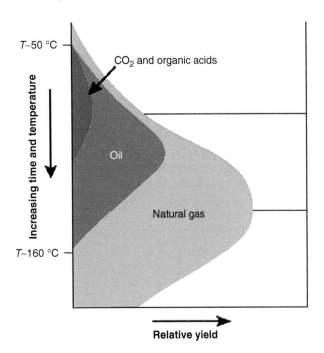

Fig. 5.4—Illustrating the conversion sequence of organic material to oil and/or gas as a function of time and temperature. Adopted from Boyer et al. (2006).

specific byproducts during continued burial as temperature and pressure increase and thermocatalytic conversion occurs.

Recall Type I and II categories generally produce oil, while Type III produces gas. Type IV are not widespread and generally do not contain hydrocarbons.

Heat and pressure drive off water and increase the carbon dioxide (CO_2) content of kerogen. Continued burial further degrades the long-string, high molecular weight, molecules to form hydrocarbon compounds of increased hydrogen content, such as methane. Vitrinite forms at the last part in the maturation process.

Fig. 5.3 illustrates the utility of the van Krevelen plot to determine the viability of a particular prospect.

Maturation and Generation of Hydrogen. Maturation gradually changes high molecular weight kerogen into smaller-sized molecules, or into methane gas. Maturation progresses in three stages:

1. Early burial and low-temperature oxidation of organic material.
2. Burial increases formation pressure and temperature; increased temperature and pressure cause cracking of the larger molecules; kerogen is gradually transformed into hydrocarbon and nonhydrocarbon gases.
   Chemical compounds in the shale and organic material begin to degrade; heavier-molecular-weight hydrocarbon compounds transform to smaller structures to form oil and gas molecules.
3. Continued maturation eventually transforms kerogen to carbon (C) and hydrogen (H) to lignite or coal.

Fig. 5.4 shows a cartoon of the maturation windows defining the presence or absence of oil or gas fluids. The figure illustrates the maturation process requires not only the correct type of organic material, but also formation within restricted temperature range. Increasing temperatures continue to alter the makeup kerogen.

The oil and gas potential of a source rock depends on the following characteristics. Analyses of cores and logs provide the necessary information.

- The source bed must contain a minimum organic content to be a viable candidate. Total organic content (TOC) is a measure of this parameter and can be determined from laboratory studies of well logs.
- Minimum cutoff for shale is near 2.0% organic material. Boyer et al. (2006) state, "2% to 4% (TOC) is usually considered good while 0.5 to 1.0% is poor."
- Thermal maturation of a sample is related to a maximum temperature peak (Tmax) that can be measured in the laboratory. Vitrinite reflectance indicates the increasing thermal alteration of lignins to an end-stage

form. Maturation values change as a function of the total organic content of the sample. The percent of vitrinite reflectance (Ro) relates and identifies the maturation history of the sample as a function of time and temperature. Lighter hydrocarbons will no longer remain in the organic material when reduced to vitrinite.
- Depositional environment
- The type of kerogen, with Type I or II preferred
- Level of thermal maturity
- In conclusion, the following summarizes the critical characteristics a viable very low permeability source rock shale accumulation needs to become a candidate for additional interest.
- Reduced oil and water saturations; high °API gravity, low viscosity oil with attendant gas.
- High gas saturation usually indicates a higher than normal relative permeability to gas. This is an indicator of the possibility that the sedimentary section contains silt and very-fine-to fine-grain-sized material. The likelihood for increased drainage area and producing rate within the area of interest increases.
- Moderate to high organic content (TOC) with a high degree of maturity. The correct type and quantity of organic material must be present for commercial generation of oil and gas.
- Well logs generally exhibit high radioactivity and resistivity when in the presence of oil and gas in organic shales.

Depletion Model

Increasing efficiency in drilling horizontal wells has a direct impact on prospect economics. Shale wells are drilled laterally up to several thousand feet through the prospective section of the reservoir to encounter as much potential pay as possible. Several fracture stages are hydraulically induced along the borehole to create a high conductivity plane within the reservoir and produce maximum reserves at minimal cost.

Figure 5.5 illustrates a completion and depletion model with the drainage area outline arranged as a function of the aspect ratio (Poston and Poe 2008). The drainage area aspect ratio (AR) is defined as the ratio of the directional extent of the rectangular drainage area in the x-and y-direction.

$$AR = \frac{x_e}{y_e}. \quad\quad\quad\quad\quad\quad\quad\quad\quad\quad\quad\quad\quad\quad\quad\quad\quad\quad\quad\quad (5.1)$$

The drainage area boundary of a multiply fractured horizontal well is assumed to be rectangular initially. Expansion of the drainage limits causes the outline to approach a more oval shape. The aspect ratio (AR) characterizes the shape in the x- and y-direction. A locus of points connecting the extent of the fracture wings defines the drainage area—i.e., stimulated reservoir volume (SRV). Reservoir fluids migrate from the reservoir matrix between the induced vertical fractures to the vertical fractures, and then through the vertical fractures to the horizontal wellbore. The low permeability matrix rock located between the vertical fractures within the drainage area

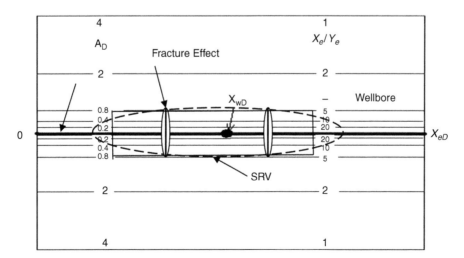

Fig. 5.5—The aspect ratio for the very-low-permeability-formation case, when the SRV is represented by the drainage boundary trending outward in the *y*-direction and wellbore length along the *x*-axis. In this case, the dashed line represents the SRV or approximate drainage area. The SRV is a function of matrix permeability and width of the multiple fracture wing lengths. In this case, permeability is in the nanodarcy range.

drained by the vertical fractures is often defined as the SRV. The pressure gradient in the SRV region has been reported by Newsham and Rushing (2001) to be possibly as high as 3 psi/ft.

Recall that the system characteristic length (L_c) for a vertically fractured well (and for a multiply fractured horizontal well) is equal to the average effective fracture half-length (X_f)—i.e., the wing length of the vertical fracture. Therefore, the width of the drainage area of a multiply fractured horizontal well completed in a low permeability reservoir is roughly equal to or slightly larger than the average total fracture length ($2L_c$). The length of the rectangular drainage area is approximately equal to the product of the number of vertical fractures and the distance between adjacent fractures ($n_f d_f$). The drainage area of a multiply fractured horizontal well (approximately equal to the SRV) is given by

$$A = 2n_f d_f X_f. \qquad (5.2)$$

Flow System. The flow system of a multiply fractured horizontal well is characterized by the following:

- The horizontal well provides a conduit to the surface and operates in a linear flow mode.
- Fracture network—Multiple finite conductivity fractures intersect along the wellbore (typically normal to the horizontal well axis). Intervening matrix blocks are the source of flow to the fractures, which then feed in to the wellbore. Initially stabilized drainage can be approximated in the shape of an ellipse surrounding each fracture before eventually interfering with adjacent vertical fractures.
- Matrix block—Low permeability reservoir matrix lying between the hydraulic fractures in a multiply fractured horizontal well.
- The volume lying within the drainage volume is designated as the SRV. The SRV might behave as a rectangular or ellipsoid reservoir volume with apparent no-flow boundaries in extremely low permeability formations.
- The unstimulated reservoir matrix located outside of the drained volume (outside of the no-flow boundary) is often of such low permeability that communication with the SRV can be assumed as nonexistent. In some cases, however, the formation permeability might be sufficient for the well's drainage area boundary to gradually expand. Contribution to the total oil production might be quite significant when such a "sweet spot" is encountered.

Transitioning from Infinite Acting to Boundary Acting Flow. The reciprocal rate-transient ($1/q$ vs. t) and pressure-transient (p_{ws} vs. t) solutions developed for a vertically fractured horizontal well exhibit similar early transient performance. Bilinear, linear, and pseudoradial flow behavior might be present in the history. Note that these relationships expressed in Eq. 5.3 for any well type are valid for all time values (Poe 2003). The discussion in Chapter 3 showed that the value of the reciprocal of the rate-transient solution at any given value of time will always be greater than or equal to the dimensionless wellbore pressure of the corresponding pressure-transient solution at the same time level.

$$\frac{1}{q_{wD}(t_D)} \geq p_{wD}(t_D). \qquad (5.3)$$

The transient performance of a multiply fractured horizontal well exhibiting linear flow behavior can be characterized with a graphical analysis of the reciprocal flow rate vs. the square root of time plot ($1/q$ vs. $\sqrt{t}$). The solution related to the vertically fractured well performance was briefly discussed in Chapter 4.

Wattenbarger et al.(1998) reported that the reciprocal flow rate vs. square root of time ($1/q$ vs. $\sqrt{t}$) plot should theoretically initiate at the origin. However, in practice the curve often begins at some positive value for the ordinate-axis intercept and the curve eventually trends to the expected straight line. This phenomenon was attributed by Wattenbarger et al.(1998) to formation damage skin effects.

An explanation for this deviation in the intercept on the linear flow diagnostic analysis could also be that the linear flow fracture production behavior might actually result from that of a moderate (finite) conductivity fracture. This exhibits pseudo-linear flow behavior (linear flow but with a finite conductivity component), instead of an infinite conductivity fracture corresponding to a negligible fracture conductivity component. The pressure transient behavior of a vertical fracture that exhibits pseudo-linear flow has been discussed by Cinco-Ley and Meng (1988). The corresponding reciprocal rate-transient behavior also exhibits a similar deviation in the intercept in the linear flow diagnostic analysis with pseudolinear flow behavior.

Value of b-Exponent. The relationship between pseudo production time (t_p) commonly used in production decline curve analyses and actual production time (t) can be expressed in the dimensionless and dimensional form shown in Eq. 5.4. This relationship can be expressed in terms of either rate-transient or decline curve analysis variables when the appropriate Arps (1945) decline model is fit to the rate-transient response. The relationship given in Eq. 5.4 derives directly from the relationships presented previously in Chapter 3.

$$\frac{t_{mb}}{t} = \frac{t_{Dmb}}{t_D} = \frac{Q_{pD}}{q_{wD}t_D} = \frac{Q_{pDd}}{q_{Dd}t_{Dd}} \quad\quad\quad\quad\quad\quad\quad\quad\quad\quad\quad\quad\quad\quad\quad\quad\quad\quad (5.4)$$

The specific relationship between the pseudo production time and field production time is given by Eq. 5.5 when the general hyperbolic decline model is applicable for matching the well's rate-transient response.

$$\frac{t_{mb}}{t} = \frac{(1+bt_{Dd})^{\frac{1}{b}}}{(1-b)t_{Dd}}\left[1 - \frac{1}{(1+bt_{Dd})^{\frac{1-b}{b}}}\right] = \frac{(1+bD_it)^{\frac{1}{b}}}{(1-b)D_it}\left[1 - \frac{1}{(1+bD_it)^{\frac{1-b}{b}}}\right] \quad\quad\quad (5.5)$$

This relationship reduces to that given by Eq. 5.6 when the Arps (1945) exponential decline model matches rate-transient well behavior.

$$\frac{t_{mb}}{t} = \frac{1-e^{-t_{Dd}}}{t_{Dd}e^{-t_{Dd}}} = \frac{1-e^{-Dt}}{Dte^{-Dt}} \quad\quad\quad\quad\quad\quad\quad\quad\quad\quad\quad\quad\quad\quad\quad\quad\quad\quad\quad (5.6)$$

The corresponding harmonic decline model relationship between the pseudo production time and production time is given by Eq. 5.7.

$$\frac{t_{mb}}{t} = \frac{(1+t_{Dd})\ln(1+t_{Dd})}{t_{Dd}} = \frac{(1+D_it)\ln(1+D_it)}{D_it} \quad\quad\quad\quad\quad\quad\quad\quad\quad\quad\quad (5.7)$$

Studies conducted on the effect of the Arps (1945) b-exponent value on the shape of the (t_{mb}/t vs. t) plot are shown in **Fig. 5.6.** The curves range from exponential ($b = 0$) decline behavior to values matching infinite-acting transient flow conditions with the hyperbolic decline model and large exponent values ($b = 20$). Note the dramatic change in shape ranging from exponential ($b = 0$) to completely transient ($b > 1$). The shape of the curve defines the overall flow system. Note that the shape of the decline exponent on a well plot could reinforce or dispel whether boundary-dominated flow is observed in the well production behavior.

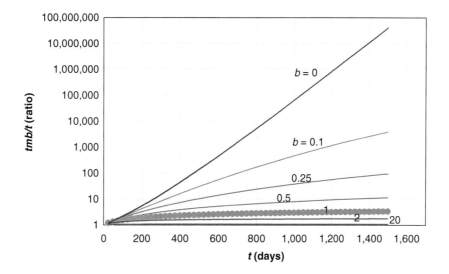

Fig. 5.6—Effect of Arps (1945) decline exponent (*b*) on the relationship between the ratio of material balance (t_{mb}) time to real time (t) and real time. The exponential (*b* = 0) curve is slightly concave upward while the remaining curves represent the increasing dominance of transient flow regimes bending the curve concave downward.

The implications of Fig. 5.6 might be summarized as follows:

- Exponential ($b = 0$)—A rapidly increasing $\left(\frac{t_{mb}}{t}\right)$ ratio. Well operates under stabilized flow conditions, indicating some combination of good communication between the wellbore and fracture system, and slightly greater than normal matrix permeability. In other words, the drainage volume (SRV) is constantly expanding.
- Hyperbolic ($0 < b < 1$)—Progression of the (b-exponent) causes the slope of the plot to rapidly decrease. Low rock matrix permeability limits the ability of the SRV to expand and gradually act as a no-flow boundary. Ilk et al. (2010) applied simulation studies to show a similar response—the Arps curves tending to flatten as a function of increasing decline exponent (b) value. The flow system ranges in quality from predominantly pseudoradial to mainly linear, according to the location of the curve between the upper and lower limits of the range.
- Transient ($b > 1$)—Flow indicating fracture unloading with little inflow effect from the surrounding matrix or a very high permeability matrix. Fig. 5.4 shows that the curves are very flat, almost horizontal for $b > 1$. For $b = 20$, the curve has a value that is only slightly larger than unity. This observation explains why the pseudo production time function generally works well for the analysis of transient production data of vertically fractured wells and, horizontal and multiply fractured horizontal wells.

The fundamental relationship between the pseudo production time function and actual production time is given in the form of an inequality by Poe (2003). In dimensional and dimensionless form, the pseudo production time function is greater than or equal to the corresponding production time. The ratio of the pseudo production time function to the production time is therefore greater than or equal to unity.

$$t_{Dmb}(t) \geq t_D(t). \quad\quad\quad\quad\quad\quad\quad\quad\quad\quad\quad\quad\quad\quad\quad\quad\quad\quad\quad\quad\quad\quad\quad\quad (5.8)$$

Conventional Plotting Methods. Conventional plotting methods are given here:

- Try to identify the boundary dominated flow portion of the production, if present.
- Apply semilog flow rate vs. time and flow rate vs. cumulative recovery plots to predict estimated ultimate recovery (EUR).
- Usually assume maximum producing time lasts no longer than 7 to 10 years to EUR unless historical production performance justifies economic production forecasts of greater duration.
- Apply no "b value" > 1 for cumulative recovery prediction of EUR. Early transient performance of acidized, fractured, horizontal, or multiply fractured horizontal wells might exhibit production behavior with decline exponent values that exceed unity, but cumulative recovery predictions of EUR should never be made with b values greater than unity.

Tubinghead pressures (FTHP) can be transformed to flowing bottomhole pressures (FBHP) to collate with the production history in order to normalize performance. Normalizing flow rate accounts for flowing pressure changes that might occur during transient conditions. Plot the pressure-drop-normalized production flow rate ($q/\Delta p$) vs. t_{mb} (pseudo production time).

Normalizing Curves.

- Log q vs. log t_m or if both sides are normalized $\left(\log\left(\frac{q}{p_i - p_{wf}}\right) \text{ vs. } \log t_{mb}\right)$. This action can result in misinterpretation
if the production history is not reviewed concurrently. This plot helps to Identify flow regimes (transient-half slope and boundary dominated flow (BDF)-1 slope).
- Linear flow $\left(\frac{p_i - p_{wf}}{q} \text{ vs. } \sqrt{t}\right)$ predominates with shale gas and should be exhibited as half-slope behavior during transient flow. Remember that damage skin effects (S) might result in the half-slope behavior being obscured in the well performance.

Shale Well Examples

Example analyses of the production performance of oil and gas wells are used to demonstrate the evaluation techniques used to quantify the well completion effectiveness and reservoir properties for multiply fractured horizontal wells that are completed in low permeability shale reservoirs. The examples chosen are multiply fractured horizontal wells located in south Texas. These wells initially unload large volumes of water before being placed on continuous production. The large volumes of water that are produced initially are primarily a result of the large

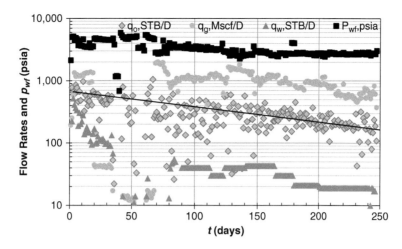

Fig. 5.7—Hixon oil well production history (raw data). Production is available for 240 days of the well history. p_i = 5,700 psi and $(GOR)_i$ = 3,300 scf/STB.

volumes of water used in the hydraulic fracturing treatments of the wells, but the wells also produce measurable quantities of formation water.

The production history of a multiply fractured horizontal oil well is presented in **Fig. 5.7.** The oil, gas, and water flow rates and the FBHP history each tend to indicate linear decline trends with increasing production time. The initial reservoir pressure was determined to be approximately 5,700 psi and the initial producing gas/oil ratio GOR is approximately 3,300 scf/STB.

The Hixon Well is a particularly good example for applying rate and pressure normalizing techniques to smooth apparently erratic data to discernible trends.

Analysis of Fig. 5.7 discloses the flowing characteristics.

- Well performance history is divided into three segments:
  1. 0 to 70 days—completion problems and unloading fracture water.
  2. 70 to 170 days—producing excessive water volume; continued pressure drawdown has caused GOR to decline.
  3. 170 to 250 days with a relatively constant FBHP (p_{wf}) because of decreased water volumes.
- Production data scatter clouds the interpretation of performance.
- Exponential decline where D = 160%/yr; need verification from other plots to determine boundary-dominated flow (BDF).
- Unexplained well history between 35 and 80 days; this is probably a result of well problems.
- Pronounced decline in water production after 180 days.

The t_{mb}/t ratio in **Fig. 5.8** seems to indicate the Hixon well never enters the boundary dominated stage. This characteristic could also be caused by extreme data scatter.

Compare Raw and Smoothed Production and Time Data. Time- and production-normalizing calculations were conducted in an attempt to reduce data scatter and increase the quality of the predictions. **Fig. 5.9** presents the semilog normalized production rate vs. time plot with the raw data. Compare the figure to the normalized production rate vs. pseudo production time (material balance time) data in **Fig. 5.10.**

To define the transient and boundary dominated segments better, the raw and normalized field data were calculated and plotted in **Figs. 5.11** and **5.12.**

Conclusion. The production decline behavior still appears to be transient (infinite-acting), with 500 days of continuous production. At the end of the available production history, we are unable to forecast reserves reliably with decline curve analysis techniques. **Fig. 5.13** shows that a reasonable match with an Arps (1945) hyperbolic decline model of approximately 1.0 is possible. However, the absence of any definitive boundary-dominated flow behavior results in a lack of uniqueness in the decline curve analyses.

Applying decline curve analysis to establish definitive reservoir and well properties must include some early transient (infinite-acting) and also stabilized (boundary-dominated) flow conditions. Without sufficient transient and boundary-dominated flow behavior to uniquely characterize a multiply fractured well's performance, the only viable approach for the well performance with transient flow data is with specialized transient production performance diagnostics and production history matching techniques.

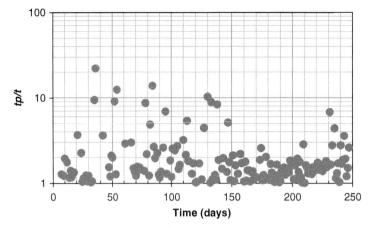

Fig. 5.8—No definitive trend in the $\left(\dfrac{t_p}{t}\right)$ ratio vs. time plot. Could an increasing trend be drawn on this graph? If so, that would indicate that the Hixon well remains in the transient state throughout its production history.

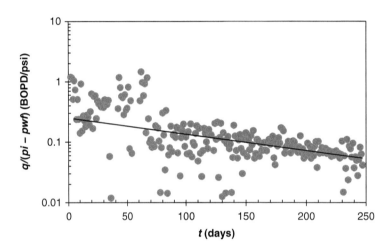

Fig. 5.9—Pressure drop normalized production flow rate vs. time plot. Apparent linear trend after approximately 80 days.

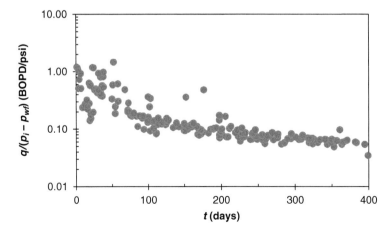

Fig. 5.10—The pressure drop normalized flow rate vs. pseudo production time plot dramatically smooths the noisy well performance data. No indication of unit slope (boundary dominated flow), but a good linear trend is observed. An apparent hyperbolic decline?

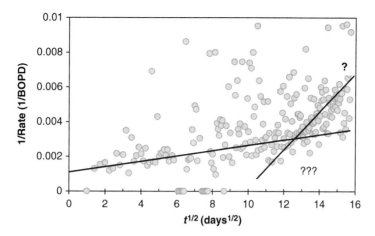

Fig. 5.11—Reciprocal flow rate vs. square root of time. The plotted raw data resulted in pronounced scatter. Are any straight line interpretations possible or probable? A possible intersection of the two flow regimes at $\sqrt{t}$ = 13, time = 169 days?

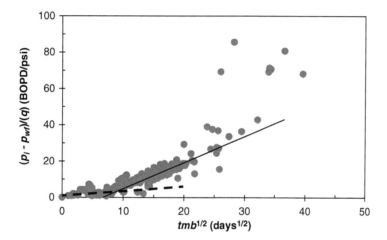

Fig. 5.12—Normalized flow rate and square root of pseudo production time smoothed data. An intersection between transient and boundary-dominated flow conditions occurs at $\sqrt{t}$ =10, t = 100 days. Unfortunately, the intersection coincides with the end of well problems shown in Fig. 5.7. What can one say approximately the data after $\sqrt{t}$ = 25 months?

Outlining Productive Limits. Anderson et al. (2010) defined a linear flow parameter (LFP) for the analysis of production performance for multiply fractured horizontal wells as the product of the matrix/fracture interface area and the square root of effective permeability of the matrix in the SRV. The linear flow parameter might be determined from the slope (m) of the line in a constant-pressure-solution graphical analysis. For a gas reservoir analysis, with $1/q_g$ vs. $\sqrt{t}$, the linear flow parameter is given by

$$LFP = A\sqrt{k} = \frac{630.8T}{m\sqrt{\phi\mu_{gi}c_{ti}}}. \quad\quad\quad\quad\quad\quad\quad\quad\quad\quad\quad\quad\quad\quad\quad\quad\quad\quad\quad\quad (5.9)$$

The area in the equation is defined as the "matrix/fracture interface area" through that the flow enters the fracture system. It is the total fracture face area of all the opened/reopened natural and hydraulic fractures accepting flow from the matrix. The product of this area and the length of the linear flow path can be interpreted as the SRV.

The same SRV can be also represented as the product of drainage area and average formation thickness. However, it is important to notice that the matrix/fracture interface area and the drainage area are not the same. Physically, they belong to surfaces perpendicular to each other. Numerically, the matrix/fracture interface area is usually much larger than the drainage area, because the latter's multiplier (the linear flow path in the matrix) is much smaller than the multiplier of the drainage area (the average formation thickness).

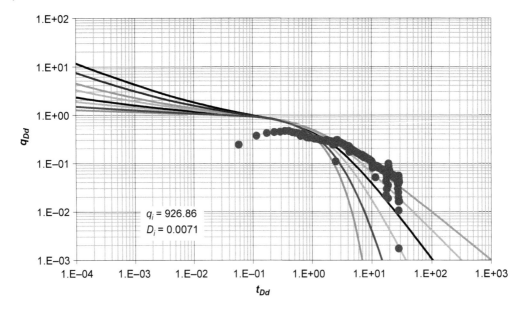

Fig. 5.13—Attempted match on Fetkovich type curve. Is $b > 1$? Match on last half of data appears to follow a decline exponent (b) stem that is slightly greater than unity. However, the transient curve does not approach any of the Arps transient curves. Do you have confidence in this prediction? Adapted from Huddleston (1991).

The slope (m) of the ($1/q$ vs $\sqrt{t}$.) plot is directly proportional to the product of the square root of the matrix effective permeability. This is also true of the total matrix/fracture interface area receiving flow from the matrix of the multiply fractured horizontal well when opened to production. In general, the steeper transient line slopes reflect a greater disparity in flow capacity between the fracture system and (SRV) matrix volume.

Recall that the intercept of the ($1/q$ vs. $\sqrt{t}$) plot should pass through the origin of the graphical analysis, as noted by Wattenbarger et al. (1998). However, it is generally considered that a positive intercept reflects the introduction of a skin effect at the fracture/matrix interface. Increased water saturation in the matrix at the fracture faces is caused by the hydraulic-fracturing-fluid leakoff, that can introduce this type of damage skin effect in the form of water blockage.

The positive intercept value can be a function of

- Matrix permeability and capillary pressure
- Effect of treating the hydraulic fracture water with surface active agents.
- The ability to reduce the induced saturation value by flowing the well back at sufficiently high rates to reduce water imbibition deeper into the reservoir matrix.

In summary, the greater the difference between the intercept and the origin in the reciprocal flow rate vs. square root of time ($1/q$ vs. $\sqrt{t}$) analysis, the greater the effect of formation damage on the well's production rate.

Regional Analysis Example

A single lease located in the south Texas Eagle Ford Shale was selected to illustrate the process of analyzing the production performance of adjacent multiply fractured horizontal "shale" wells to identify areas of interest and disinterest.

Five plotting methods were previously discussed.

1. Performance plot—Oil production rate (BOPD), water [water/oil ratio (WOR)], and gas/oil ratio (GOR) rates vs. time, to review production history and form a picture of the performance character of the well.
2. Semilog flow rate vs. time (q_o vs. t)—Are there discernible producing segments? Estimate reserves if a linear behavior is exhibited and is boundary-dominated flow.
3. Log q vs. log t—If possible, try to identify transient and boundary-dominated flow regimes.
4. Semilog (t_{mb}/t) vs. time—To help define flow type.
5. $1/q$ vs.$\sqrt{t}$ —Aids in defining transient and boundary-dominated flow for a linear system and if identified, the extent of formation damage.

Example. Fig. 5.14 shows the locations and directional orientations of the horizontal wells that are completed in the subject lease in the Eagle Ford Shale.

The monthly production for the wells was obtained from a public information service on an individual well basis. The well production data were plotted using the five previously discussed coordinate systems.

Well EW111 provides an example of the production behavior displaying both transient and boundary dominated flow conditions and are shown in **Figs. 5.15.** through **5.18**. Production data were plotted using four graphical interpretations. Refer to **Figs 5.15 through 5.18**.

Interpretation. From Figs 5.15 through 5.18, it can be concluded that

D = 25.8%/yr, production to date = 43,800 BO, remaining recoverable production = 5,110 BO, boundary regime ≈ 200 – 225 days, b = 0.005 1/BOPD, m = 0.00085 (BOPD- days $^{1/2}$)$^{-1}$

There is some indication of formation damage skin effect.

Each well was subjected to this plotting and analysis procedure. Results are summarized in **Table 5.2.** A portion of the resulting table is shown below.

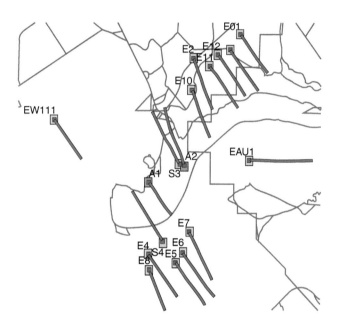

Fig. 5.14—Horizontal well completions in south Texas Eagle Ford Shale lease.

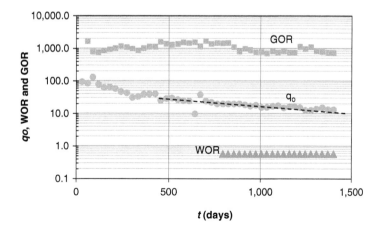

Fig. 5.15—Performance plot of flow rate, WOR, GOR, and semilog analysis of flow rate. Well-defined straight-line depletion, constant GOR, and water production.

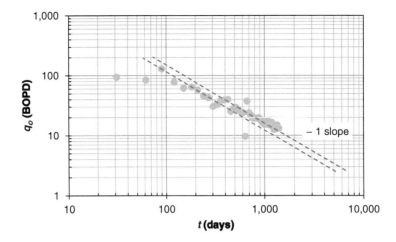

Fig. 5.16—Log production rate vs. log time analysis. Good fit signifying unit slope began at approximately 200 to 400 days.

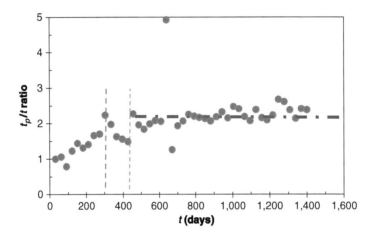

Fig. 5.17—Time ratio plot shows a break over the 300- to 400- day window.

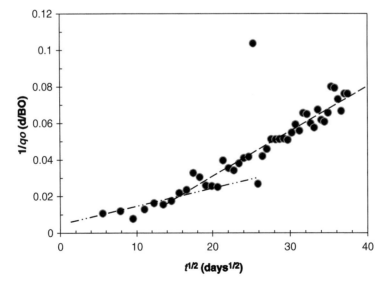

Fig. 5.18—Reciprocal flow rate vs. square root of time analysis. Transitions from transient to BDF regime over 225 to 300 day interval.

Transient and boundary-dominated flow conditions are apparent in two of the 17 lease wells, with two additional wells that seem to have begun displaying transition between transient and boundary-dominated flow. Data variability or the lack of a linear trend often precluded making similar conclusions for the remainder of the wells. **Fig. 5.19** seemingly indicates four major areas displaying similar production characteristics located within the lease area. These producing areas are:

1. North area—marginal reserves potential.
2. EW111—mediocre well located at northwest periphery.
3. EAU1—single high value well located at the southeast periphery. Is there appraisal potential?
4. Southern portion of lease—developed by Wells S3, A1, E4, S4, and E8. Well E4 was always a poor producer. Is this an indication of completion problems? Southern boundary defined by Wells E5, E6, and E7.

Note the two outliers. Well A1 probably suffered casing collapse or sanded up at approximately 1,200 days. Well E4 seems to have suffered an ineffective completion and has always been a poor producer. Could reserves be over or underestimated?

Well	N_p Reserves MBO	Identify Transient	Bound $D = \%$	Analysis Characteristics
PR11	31.9 + 4.6 = 37		78.8	Highly variable production data Ineffective fracturing job?
PR12	24.8 + 16.5 = 41		28.7	Transient throughout life
PRSAU	120 + 191.6 = 312		28.0	High rate well
EW111	43.8 + 5.1 = 49	$b = 0.005$ $m = 0.00085$	25.8	Well-defined history Well-defined exponential decline
PR3	85.9 + 86.0 = 170.0		?	Oscillating horizontally ≈100 BOPD. Still infinite acting?
PR4b	58.8 + 58.8			Oscillating horizontally ≈100 BOPD Still infinite acting?

Table 5.2—Well Analysis Summary. Mbo = 1,000 Bo.

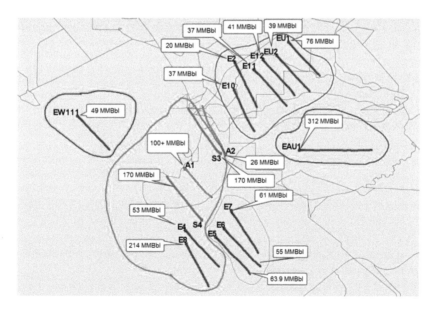

Fig. 5.19—Eagle Ford Shale multiply fractured horizontal wells grouped according to their EURs.

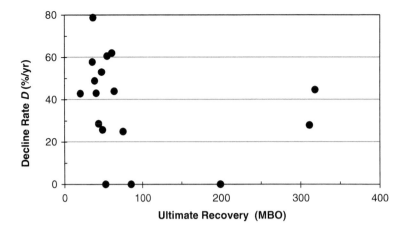

Fig. 5.20—The plot defines the decline rates as a function of ultimate recovery on a well basis. Lease well decline rates range from good to bad, with most of the wells ranging from 25 to 60%. Unfortunately, the decline rates for two of the good wells lie within this same range. MBO = 1,000 BO.

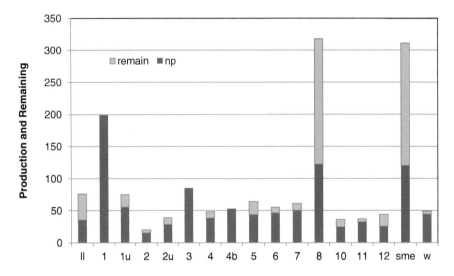

Fig. 5.21—Relationship between credited production plus expected future performance by well. Wells EAU and E8 certainly require further review. Well A1 was a good producer, why no additional reserves?

Conclusions and Recommendations. Graphical analyses and collation of results revealed these follow-up questions for the Eagle Ford study area:
- Would you recommend offsetting Well EAU1 in this lease area?
- Review workover potential of the two wells located in the southern area and a possible extension of the current productive northwest limit.
- In both cases, geological studies must be incorporated into the analysis for a more in-depth study (Willis and Tutuncu 2014).

Early Time Behavior of Fractured Horizontal Wells. The early production performance of a fractured horizontal well can be evaluated using transient models such as the ones developed by Chen and Raghavan (1997) or Ozkan et al. (2011). At early production time, the effects of external boundaries do not influence wellbore transient response of a fractured horizontal well. Performance during this early production time essentially corresponds to that of a well located in an infinitely large reservoir.

The early rate-transient behavior of a multiply fractured horizontal well in a single-porosity reservoir with (n_f) identical, uniform, equally spaced vertical fractures can be reasonably well-organized into a single family of production rate decline curves, correlated in terms of the dimensionless fracture conductivity.

A family of production decline curves of this type would be required to adequately include the effects of dual-porosity-reservoir parameters in a dual-porosity system. However, as an example, the production rate decline behavior of a multiply fractured horizontal well with nf identical, equally spaced vertical fractures in an infinite-acting, single-porosity, isotropic reservoir is presented in **Fig. 5.22**. Except at very early transient times, wellbore standoff, wellbore radius, and average fracture width generally have only a limited effect on longer term rate-transient behavior of a multiply fractured horizontal well. A range of selected values of dimensionless fracture conductivity covering a range from low dimensionless conductivity ($C_{fD} = 1$) to high dimensionless conductivity ($C_{fD} = 500$) are included in Fig. 5.22.

Single-porosity-reservoir models are not generally applicable for characterizing rate-transient performance of a fractured horizontal well located in a naturally fractured or contrasting permeability laminated dual-porosity, low-permeability unconventional reservoir. Therefore, it is generally preferred to generate a customized set of rate-transient decline curves using the Chen and Raghavan (1997) model or with the tri-linear model of Ozkan et al. (2011) for analyzing the dual-porosity case.

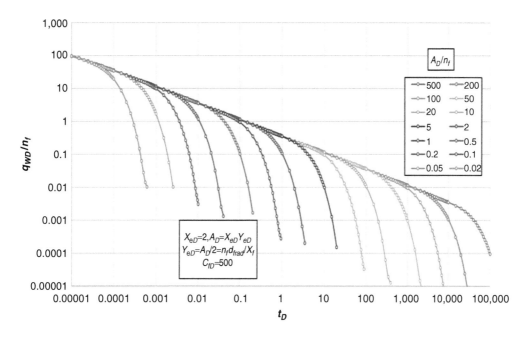

Fig. 5.22—Rate-transient behavior of a fractured horizontal well in a single-porosity unconventional reservoir where $C_{fD} = 500$.

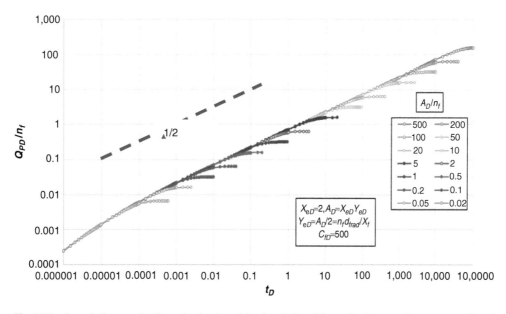

Fig. 5.23—Cumulative production of a fractured horizontal well in a single-porosity unconventional reservoir where $C_{fD} = 500$.

It has been found in various reservoir simulation studies of low-permeability unconventional reservoirs that the reservoir volume drained is effectively limited to an area defined by the average effective length of the vertical fractures intersected by the horizontal well, and by the overall distance between the drainage extent of the outermost two vertical fractures of the horizontal well. Raghavan and Joshi (1993) and Chen and Raghavan (2013) have considered this effect of the reservoir drainage in the direction along the horizontal well axis.

The rate-transient behavior of a multiply fractured horizontal well completed in a low-permeability, unconventional reservoir is not easily correlated into a single generalized rate-transient production decline curve. An approximation of the rectangular SRV dimensions and reservoir drainage area must be considered.

The rate-transient performance of multiply fractured horizontal wells located in low-permeability unconventional reservoirs is best evaluated with the multiply fractured horizontal well performance model. Specific reservoir and well completion properties of interest can then be included for this case. Poe* (2018) has developed rate and cumulative recovery curve for a range of hydraulic fracture dimensionless conductivities, (C_{fD}=500, 200, 100, 50, 20, 10, and 5), which can be obtained from that author upon request.

In each of these figures, the flow rate and cumulative production values have been normalized by the number of vertical fractures intersecting the horizontal wellbore to further generalize the rate-transient performance. The drainage area of a multiply fractured horizontal well is a function of the number of vertical fractures intersected by the horizontal well and also of the average distance between adjacent fractures. Multiply fractured horizontal wells are commonly selectively completed with perforation clusters along a cemented casing or liner, and then fracture stimulation treatments are performed in stages to obtain effective stimulation coverage. The depth of investigation from a fracture and the production time required for interference to occur between adjacent fractures must be considered.

Fig. 5.22 is an example of one of the rate vs. time plots, and Fig. 5.23 is an example of a cumulative production vs. time plot where C_{Df} = 500. Note that the utility of the plots has been expanded by including the number of fractures included in the production summary.

---

*Poe, B. D. Jr. 2018. Personal Communication.

Chapter 6

Type Curves

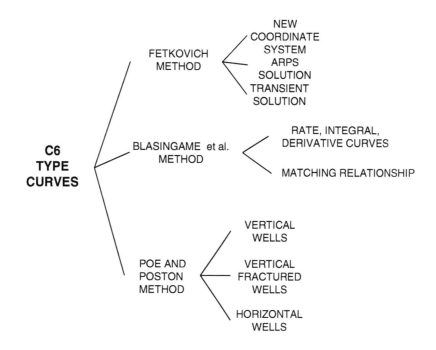

Introduction

Production decline curves have been developed and used for analyzing the production performance of various types of well completions and reservoirs. Fetkovich (1980) presented a landmark paper in which the transient and boundary-dominated flow behavior for an unfractured vertical well located in a closed cylindrical reservoir was melded into a single composite production decline solution. Theoretically, one can evaluate production decline behavior of a well over the entire transient and bounded flow production history. Fetkovich (1980) developed graphical scaling parameters to smooth the transition between infinite-acting transient and boundary-dominated flow conditions at later transient time. The Arps (1945) empirical production decline solutions represent boundary dominated flow conditions.

Uniqueness issues can arise when evaluating production behavior exhibiting only infinite-acting transient, or even some early transitional, flow behavior. Reservoir drainage area cannot be realistically determined because boundary-dominated flow performance when the effects of the reservoir limits (drainage area size, shape, and outer boundary type) are not exhibited.

Two notable analysis methodologies have been developed for production decline curve analyses since Fetkovich (1980) introduced the composite decline curve methodology. Later methods attempted to localize early-time boundary affects to forecast performance. These methods are

- Doublet and Blasingame (1995, 1996) applied integral and integral-derivative function transformations in graphical production decline analyses to aid interpretation.
- Poe and Poston (2010) applied a well and reservoir model- approach to reduce uniqueness problems resulting from the lack of boundary-dominated flow data. A graphical analysis technique was developed where transient behavior is simultaneously characterized by the flow rate decline and pseudo-production time functions.

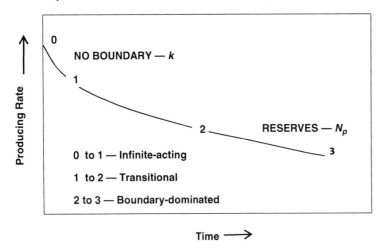

Fig. 6.1—General production rate decline flow regimes. Adapted from Matthews and Russell (1967).

Fig. 6.1 illustrates the progression of drainage histories commonly observed during the life of a well. An infinite-acting reservoir transient flow period that is not affected by reservoir boundaries operates initially. A transition flow period follows during which the effects of the reservoir limits begin to increasingly affect the well's transient production behavior. Finally, a boundary-dominated flow regime will be exhibited in the production performance in which the effect of reservoir boundaries governs well transient performance.

During the 0 to 1 flow regime, the well typically begins producing under transient flow conditions. Express solution in terms of the exponential integral function for the unfractured vertical well case. Other well types might exhibit transient behavior including spherical, bilinear, or linear flow, some which might exist for extraordinarily long periods of time when producing from very low permeability reservoirs, particularly in systems that have an elongated source/sink. Examples of these types of wells include vertically fractured wells, horizontal drainholes, and multiply fractured horizontal wellbores.

During the 1 to 2 flow regime, there is a continual transition from totally infinite-acting transient flow conditions to fully boundary-dominated flow behavior. Quantitative production decline analysis techniques are not available for the transition flow regime.

During the 2 to 3 flow regime, reservoir pressure declines as a function of well cumulative production located in a closed, finite reservoir. The boundary-dominated flow behavior of a well located in a closed, finite reservoir is described in pressure-transient analyses (terminal flow rate inner boundary condition) with the classic pseudosteady-state solution of the governing diffusivity equation. The analytic solution of the flow rate decline for a well in a closed, finite reservoir corresponds to the exponential decline solution.

The Problem

We need to be able to detect and analyze the boundary effects signals in Segment 1 to Segment 2 as early as possible to be able to estimate reserves, particularly for the case of very low permeability reservoirs.

The type curve analysis approach involves matching a set of production data points to a reference solution to the diffusivity equation. The solutions have previously been developed for a set of boundary conditions assumed to correspond to those of the producing well. The reference type curve solution used in the analysis corresponds to the specific well type and reservoir type that pertain to the well whose production performance is being characterized in the analysis.

Dimensionless and field data match points are obtained once an overlay of the data curve on the reference type curve set results in an acceptable match. These match points represent the numerical difference between the reference type curve solution of the diffusivity equation and the field data curve. A word of caution though: the analyst inherently assumes that the field data and the reference type curve solution share the same applicable assumptions and boundary conditions. This might not always be true. Saying an apple is an orange does not necessarily make it an orange.

Type curves are a graphical presentation of the solution of the diffusivity equation governing fluid flow in a reservoir for a set of reference conditions such as well type, reservoir matrix type, and outer boundary conditions to compare measured production performance behavior of an active well. The inner boundary condition applied to generate type curves for use in pressure-transient (well test) analyses is the terminal flow rate (Neumann) inner boundary condition. The terminal flow rate inner boundary condition is commonly used to construct the type curve reference solutions, and is used with the superposition principle to correspond to typical well test conditions for shut-in transients of pressure-buildup or- falloff analyses.

However, the reference type curve solutions are typically constructed with the rate-transient terminal flowing pressure (Dirichlet) inner boundary condition solutions. The reason for this is that under normal production operations, somewhere in the production system there is a terminal flowing pressure inner boundary condition. After the well has been on production for a period time, or in the surface facilities with a constant separator backpressure, or constant sales-line pressure, the terminal pressure inner boundary condition may be applicable. This is because of an approximately constant flowing bottomhole pressure condition. Except for an actual shut-in well condition, somewhere in the production operating system there will generally be a point that corresponds closely to a terminal pressure inner boundary condition. A series of terminal pressure inner boundary condition steps can be applied in production analyses when a constant terminal pressure inner boundary condition cannot be exactly maintained.

The van Everdingen and Hurst (1949) constant terminal pressure solution (q_D) of the diffusivity equation can be used to construct reference type curves for an unfractured vertical well in a closed cylindrical reservoir. To express the dimensionless well flow rate solution (q_D) in terms of the dimensional parameters, we have

$$q_D = \frac{141.2 q \mu B}{kh(p_i - p_{wf})} \quad \quad (6.1)$$

Taking the logarithm of the dimensionless flow rate definition,

$$\ln q_D = \ln q + \ln \frac{141.2 \mu B}{kh(p_i - p_{wf})} \quad \quad (6.2)$$

Define dimensionless time (t_D) in the general form given in Eq. 6.3. The system characteristic length (L_c) of an unfractured vertical well is equal to the wellbore radius (r_w). For a vertically fractured well, the system characteristic length is equal to the fracture half-length (X_f), and in an unfractured horizontal drainhole the system characteristic length is equal to half of the effective length of the horizontal drainhole in the reservoir ($L_h/2$). System characteristic length of a multiply fractured horizontal well is equal to the average of the effective fracture half-lengths located in the multiple-fracture system.

$$t_D = \frac{0.00633 kt}{\phi \mu c_t L_c^2} \quad \quad (6.3)$$

The definition of the dimensionless time (t_D) can also be expanded in terms of logarithms in a similar manner.

$$\ln t_D = \ln t + \ln \frac{0.00633 k}{\phi \mu c_t L_c^2} \quad \quad (6.4)$$

The dimensional field data (flow rates and time) and corresponding dimensionless reference variables differ only by a set of constant values. This fact is the basis for all type curve analyses procedures. One simply separates the variables and constants of a solution to the diffusivity equation. The solutions are assumed to match when the curvature of both the overlay field data curve and the reference type curve agree. The difference in the match points is simply constant values dependent on reservoir and well completion properties. Match points selected for the two coordinate systems are then used to quantitatively determine the effective permeability and system characteristic length.

Fetkovich Method

The dimensionless decline flow rate (q_{Dd}) and time (t_{Dd}) variables developed by Fetkovich (1980) and used by Fetkovich and Thrasher (1979) and Fetkovich et al. (1987) are defined in Eqs. 6.5 and 6.6, respectively. Note that the system characteristic length used in these relationships follows from the Prats (1961) apparent wellbore radius (r_{wa}) concept as seen in Eq. 6.7, instead of the wellbore radius, to include steady-state skin effects in the type curve analyses. In addition, note that the constant term "½" present in these expressions differs from the constant value of "¾" traditionally associated with pressure-transient solutions for boundary-dominated flow. This change was empirically made in the analysis variables by Fetkovich (1980) to obtain a better coupling of the infinite-acting transient and Arps (1945) boundary-dominated flow stems in the composite decline curve solution.

$$q_{Dd} = \frac{141.2 q \mu B \left[\ln\left(\frac{r_e}{r_{wa}}\right) - \frac{1}{2} \right]}{kh(p_i - p_{wf})} \quad \quad (6.5)$$

$$t_{Dd} = \frac{0.01266 kt}{\phi \mu c_t r_{wa}^2 \left[\left(\frac{r_e}{r_{wa}}\right)^2 - 1 \right] \left(\frac{r_e}{r_{wa}} - \frac{1}{2}\right)} \quad \quad (6.6)$$

$$r_{wa} = r_w e^{-s} \quad \quad (6.7)$$

Evaluating the logarithm of the Fetkovich (1980) dimensionless decline time (t_{Dd}) and flow rate (q_{Dd}) functions results in expressions to form the basis for a graphical type curve analysis. These logarithmic transformations are given in Eqs. 6.8 and 6.9.

$$\ln q_{Dd} = \ln q + \ln \left\{ \frac{141.2 \mu B \left[\ln\left(\frac{r_e}{r_{wa}}\right) - \frac{1}{2} \right]}{kh(p_i - p_{wf})} \right\} \quad \quad (6.8)$$

$$\ln t_{Dd} = \ln t + \ln \left\{ \frac{0.01266k}{\phi \mu c_t r_{wa}^2 \left[\left(\frac{r_e}{r_{wa}}\right)^2 - 1\right]\left[\frac{r_e}{r_{wa}} - \frac{1}{2}\right]} \right\} \quad \quad (6.9)$$

The difference between the dimensionless reference type curve solution and the well's dimensional production data can be determined from an overlay of the logarithmic flow rate vs. time curves. Both data sets are plotted on the same log-log scale. The last term in each of Eqs. 6.8 and 6.9 represent the relationships between the dimensionless type curve solution and the dimensional production data, expressed as functions of the reservoir and well completion properties.

New Coordinate System. Fetkovich (1980) defined three new dimensionless variables for use in decline curve analyses. These are production decline analysis time (t_{Dd}), flow rate (q_{Dd}), and cumulative production (Q_{pDd}) variable transformations, developed specifically for use in graphical production decline curve analyses. The addition of an additional subscript ($_d$) to the variable names denotes variable definitions pertaining to the Fetkovich (1980) type production decline curve analyses, instead of conventional rate-transient analyses techniques. These are given as follows.

Dimensionless decline flow rate,

$$q_{Dd} = \frac{q_2}{q_i} \quad \quad (6.10)$$

Dimensionless decline time,

$$t_{Dd} = D_i t \quad \quad (6.11)$$

Dimensionless decline cumulative production,

$$Q_{pDd} = \frac{D_i Q_p}{q_i} \quad \quad (6.12)$$

Application of the type curve approach provides a means to transfer solutions to specific boundary conditions to solve practical field problems.

Fig. 6.2 shows that the dimensionless drainage radius ($r_{eD} = r_e/r_{wa}$) correlating parameter is for the infinite-acting transient stems, while the Arps (1945) decline exponent (b) is the correlating parameter for the boundary-dominated flow stems. Note that both parameters relate to the reservoir size or boundary-dominated flow behavior. Generate the transient stems of the Fetkovich (1980) composite decline type curve with the transient solution of the diffusivity equation given by van Everdingen and Hurst (1949). The Arps (1945) models define the flow rate decline behavior under boundary-dominated flow conditions.

Remarks and Limitations. The slopes of the Fetkovich (1980) depletion stems in Fig. 6.2 are very similar in the transition flow regime range ($0.1 < t_{Dd} < 2$). A unique decline curve analysis match can be achieved only if there is sufficient curvature in the field data and match curves to quantitatively identify the appropriate decline curve stem. In addition, to obtain a reasonably unique decline curve analysis match involves having at least some production performance exhibited under transient flow conditions ($t_{Dd} < 0.1$) and at least some production behavior exhibited after the onset of boundary effects. This means at least some production performance must also include

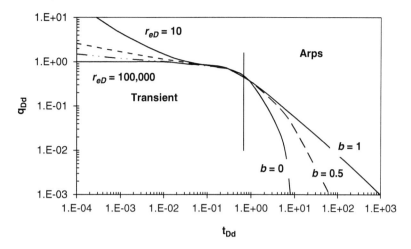

Fig. 6.2—Composite decline curve. Modified from Fetkovich (1980).

the transition flow regime ($t_{Dd} > 0.1$). This combination of transient and transition flow behavior "anchors" the overlay field data curve on the reference decline curve set.

The van Everdingen and Hurst (1949) rate-transient dimensionless time and flow rate variables were transformed by Fetkovich (1980) into a new coordinate system that is more amenable for production decline analyses. Finding the match point between the curve fit of the field data and type curve provides the bridge between the dimensionless and dimensioned solutions.

Transient flow studies have shown that matching of the transient flow data to Arps (1945) models is invalid because the rate profile during transient flow should always be concave upward. The Arps (1945) production decline models all have flow rate decline behaviors that exhibit a concave downward behavior. Transient behavior of vertical fractured, horizontal, and multiply fractured horizontal wells is commonly fit using hybrid decline curve solutions consisting of one Arps (1945) decline curve model for the transient production performance of the well. The solution is then coupled with another Arps (1945) decline model for the boundary-dominated production decline behavior.

This practice is commonly used in the evaluation of the performance of wells completed in low-permeability unconventional reservoirs, because the late time production behavior of the well might not be exhibited for a very long time and a hybrid decline curve analysis model may be considered adequate. The technique forces the curve to fit the well behavior to an empirical production decline model. Additional discussion of these hybrid models can be found in Chapter 1.

Fetkovich et al. (1987) provided additional insight into the validity of traditional decline curve analyses by stating that "...a reliable pore volume estimate cannot be obtained if the transition indicating the approach of boundary-dominated conditions is not observed." Stated again, if the effects of a reservoir or well completion parameter are not exhibited in the production performance of a well, then that parameter cannot be reliably determined by conventional production decline analysis. However, more-recent decline curve analysis techniques have been reported that can be used to help reduce some of the uncertainty in the results of a production decline curve analysis when only early transient well performance data are available.

Conclusion. The decline analysis dimensionless variables introduced by Fetkovich (1980) allow the presentation of the infinite-acting transient production decline performance and the Arps (1945) model late-time production decline behavior in a single composite production decline type curve. This concept enables the composite decline analysis type curve solution to be more readily used for production decline curve analyses.

Arps Side of Fetkovich Plot. The Arps (1945) hyperbolic production rate and time variable definitions are combined with the Fetkovich (1980) dimensionless production rate and time variable definitions to describe the production decline behavior. Note that the Arps (1945) decline functions are simply equations of a curve, while the solutions that describe the transient behavior of a well are solutions to a fluid flow equation. The two solutions can be coupled to correlate the production decline behavior of a well even though they are dissimilar functions,

The rate-transient solution for a well that is completed in a closed reservoir and the Arps (1945) decline curve models are correlated with the solution for the exponential decline curve. The boundary-dominated flow behavior of the rate-transient solution of the diffusivity equation describing production performance of a well located in a closed reservoir is like the classic exponential decline model. This model corresponds directly to the flow rate decline behavior described by the empirical Arps exponential ($b = 0$) decline curve model.

$$q_{Dd} = \frac{q_2}{q_i} = \frac{1}{(1+bD_i t)^{\frac{1}{b}}} \quad \quad \quad (6.13)$$

$$t_{Dd} = D_i t = \frac{\left(\frac{q_i}{q}\right)^b - 1}{b} \quad \quad \quad (6.14)$$

Information Obtained From the Analysis.
The information obtained from the analysis includes

- Arps (1945) decline exponent (b) or dimensionless drainage radius (r_{ed}) value.
- q_i and d_i for input into the Arps (1945) or pseudosteady-state equations.
- Extending production trend along matched type curve to estimate future production.

The Fetkovich (1980) type curves were developed to cover the spectrum of conditions expected to be encountered in practice. Note that short duration production tests will generally tend to exhibit transient production behavior, while longer term production data records could include late-time production performance data that would correspond to the Arps (1945) production decline model behavior. The dimensionless decline flow rate behavior of the Arps (1945) decline curve analysis models is illustrated in **Fig. 6.3** for the exponential, hyperbolic, and harmonic decline curve solutions.

Analysis Procedure.
Follow these steps to conduct the analysis procedure:

1. Combine the match points $(t, q)_{mp}$ and $(t_{Dd}, q_{Dd})_{mp}$ and the Fetkovich (1980) transformations to calculate the initial decline and rate coefficients.
2. Extend the production data curve downward along the fitted reference type curve solution to predict the future production.

Please note that, the initial flow rate (q_i) and decline rate (D_i) can be evaluated using the match point values of time and flow rate from the production data and reference decline curves: $q_i = \frac{q_2}{q_{Dd}}, \quad D_i = \frac{t}{t_{Dd}}$.

The decline rate is constant for the exponential ($b = 0$) decline. Therefore, the initial flow rate value is not required to complete the calculations.

3. The decline curve stem that corresponds to a harmonic decline curve is $b = 1$. The initial flow rate and decline rate coefficients must be calculated to apply the flow rate, time, and cumulative recovery relationships.
4. The decline curve stem of a hyperbolic curve has a decline exponent (b) value that varies between the exponential and harmonic decline exponent values ($0 < b < 1$) for late-time production behavior. The

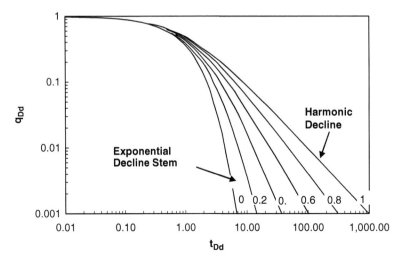

Fig. 6.3—Arps (1945) late-time production decline model behavior. Modified from Fetkovich (1980).

decline exponent will commonly exceed a value of unity when this model is applied for the analysis of transient production data. The initial flow rate and decline rate coefficients must be calculated to apply the flow rate, time, and cumulative recovery equations.

5. A match of the Arps (1945) late-time decline curves provides estimates of the initial decline rate (D_i) and initial producing rate (q_i). Insert these values back into the exponential, hyperbolic or harmonic equations to solve for the flow rate, time, and cumulative production.

The Transient Side of the Fetkovich Curves. Apply the van Everdingen and Hurst (1949) solution to evaluate the rate-transient performance of an unfractured vertical well located in a closed cylindrical reservoir. Plot the solutions for the infinite-acting transient decline stems. Early transient behavior flattens as a function of the reservoir size and late-time transient behavior collapses to a single exponential decline ($b = 0$) stem. See **Fig. 6.4.** Matching of production data only in the range $t_{Dd} > 10^{-1}$ should not be attempted because the decline curve stems are essentially of the same shape, with very little separation.

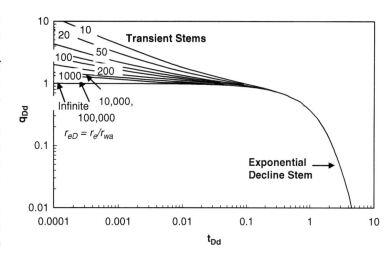

Fig. 6.4—Transient decline curve stems. Modified from Fetkovich (1980).

Rate and Time Solutions. Eq. 6.15 provides the definition for the dimensionless decline flow rate of the Fetkovich (1980) solution for constant bottom hole pressure production of a vertical unfractured well.

$$q_{Dd} = \frac{141.2 q \mu B \left[\ln\left(\frac{r_e}{r_{wa}}\right) - \frac{1}{2} \right]}{kh(p_i - p_{wf})} \quad \quad \quad (6.15)$$

Rearrange Eq. 6.15 to solve for the formation conductivity (kh), for liquid flow rates in units of STB/D.

$$kh = \frac{141.2 \mu B}{p_i - p_{wf}} \left[\ln\left(\frac{r_e}{r_{wa}}\right) - \frac{1}{2} \right] \left[\frac{q}{q_{Dd}} \right]_{mp} \quad \quad \quad (6.16)$$

Eq. 6.16 is applicable for decline curve analyses of the production performance of oil reservoirs. Eq. 6.17 shows the formation conductivity can be evaluated in a Fetkovich (1980) decline curve analysis with the appropriate expression in terms of real gas pseudo-pressure for gas reservoir analyses. The gas flow rate in this relationship is expressed in units of Mscf/d.

$$kh = \frac{1422 T}{p_p(p_i) - p_p(p_{wf})} \left[\ln\left(\frac{r_e}{r_{wa}}\right) - \frac{1}{2} \right] \left[\frac{q}{q_{Dd}} \right]_{mp} \quad \quad \quad (6.17)$$

An approximation of the real-gas pseudo-pressure potential (p_p) can be expressed in terms of p^2 when the reservoir pressure is less than 3,000 psi for low-pressure gas reservoirs. Eq. 6.18 shows the relationship for evaluating the formation conductivity from the flow rate match points in the Fetkovich (1980) decline curve analysis using the low-pressure gas reservoir p^2 formulation.

$$kh = \frac{1422 T \mu Z}{p_i^2 - p_{wf}^2} \left[\ln\left(\frac{r_e}{r_{wa}}\right) - \frac{1}{2} \right] \left[\frac{q}{q_{Dd}} \right]_{mp} \quad \quad \quad (6.18)$$

Eq. 6.19 defines the Fetkovich (1980) dimensionless decline time, with units of time are given in days.

$$t_{Dd} = \frac{0.01266 kt}{\phi \mu c_t r_{wa}^2} \frac{1}{\left[\left(\frac{r_e}{r_{wa}}\right)^2 - 1\right]\left[\ln\left(\frac{r_e}{r_{wa}}\right) - \frac{1}{2}\right]} \quad \quad \quad (6.19)$$

The definition of the dimensionless decline time (t_{Dd}) can be rearranged to develop an expression for evaluating the pore volume corresponding to the drainage area of the well in the reservoir. Eq. 6.20 shows the relationship for an oil reservoir.

$$V_p = \frac{\pi(r_e^2 - r_{wa}^2)\phi h}{5.615} = \left[\frac{\overline{B_o \mu_o}}{\mu_{oi}\overline{c_t}(p_i - p_{wf})}\right]\left(\frac{t}{t_{Dd}}\right)_{mp}\left(\frac{q}{q_{Dd}}\right)_{mp} \quad \ldots \ldots (6.20)$$

The pore volume connected with the well can be evaluated from the decline curve analysis match point values with Eq. 6.21 for gas reservoir analyses.

$$V_p = \frac{\pi(r_e^2 - r_{wa}^2)\phi h}{5.615} = \left[\frac{56.557T}{\mu_{gi}\overline{c_t}(p_p(p_i) - p_p(p_{wf}))}\right]\left(\frac{t}{t_{Dd}}\right)_{mp}\left(\frac{q}{q_{Dd}}\right)_{mp} \quad \ldots \ldots (6.21)$$

Eq. 6.22 shows the expression for a low-pressure ($p < 3{,}000$ psi) gas reservoir, and the pore volume connected to a well when evaluated using the decline curve analysis match points.

$$V_p = \frac{\pi(r_e^2 - r_{wa}^2)\phi h}{5.615} = \left[\frac{56.557T\overline{\mu_g Z}}{\mu_{gi}\overline{c_t}(p_i^2 - p_{wf}^2)}\right]\left(\frac{t}{t_{Dd}}\right)_{mp}\left(\frac{q}{q_{Dd}}\right)_{mp} \quad \ldots \ldots (6.22)$$

For most cases, the viscosity values appearing in Eqs. 6.20 through 6.22 are assumed to be similar and cancel out. Formation volume factors are evaluated at an average pressure, and the method works best when system drawdowns are small.

Required Information.
Evaluate total system compressibility with Eq. 6.23.

$$c_t = S_o c_o + S_w c_w + S_g c_g + c_f \quad \ldots \ldots (6.23)$$

Information Obtained.
The information obtained includes

- Formation conductivity (kh) or just permeability if net pay thickness (h) is known.
- Apparent wellbore radius (r_{wa}) and skin (S).
- Reservoir pore volume (V_p) and/or drainage area (A).

Analysis Procedure.
Follow the steps below to conduct the analysis procedure. Please note, Table 6.1 lists the reservoir information required to complete the calculations.

1. Extend the field production curve along the fitted type curve to predict future production.
2. Permeability-thickness product, reservoir pore volume, and apparent wellbore radius values can be calculated from the interpretation of match stem and match points.
3. Calculate the apparent wellbore radius (r_{wa}) which is a rearranged form of the pore volume calculation.

$$r_{wa} = \left\{\frac{0.01266k}{\phi\mu c_t\left[(r_{eD})_{mp}^2 - 1\right]\left[\ln(r_{eD})_{mp} - \frac{1}{2}\right]}\left(\frac{t}{t_{Dd}}\right)_{mp}\right\}^{0.5} \quad \ldots \ldots (6.24)$$

Conclusions. In conclusion,

- The curves on the transient side of the Fetkovich (1980) solution are rather flat, which tends to make interpretation much more problematic.

Reservoir rock and fluids properties			
Interval thickness (h), (ft)	Total compressibility (c_t), (psi)$^{-1}$	Wellbore radius (r_w), ft	BHFP (p_{wf}), (psi)
Porosity (ϕ), fraction	B_o or B_g (RB/STB or RB/scf)	Initial pressure (p_i), (psi)	μ_o or μ_g, cp

Table 6.1—Information required for Fetkovich decline curve analysis.

- Reservoir fluid and rock properties are required for solution with the transient decline curve analysis.
- Permeability, wellbore radius, skin, and pore volume are outcome values.

Analysis Procedure.
Follow these steps to conduct the analysis procedure:

1. Construct a log-log plot of the rate vs. time data on the same scale as the type curve.
2. Obtain match points and appropriate stem; match points are (t, q) and (t_{Dd}, q_{Dd}).
3. Select the (r_{eD}) stem signifying drainage volume if the transient figure is used, while the (b) stem is determined if the Arps (1945) late-time decline stems are applied for the interpretation.

Boundary-Dominated Example. Apply the Fetkovich (1980) type curve approach to calculate the volume of oil expected to be produced during the 36- to 48-month period from Redfish Well 2. **Table 6.2** lists recorded production, while **Fig. 6.5** shows the log rate vs. log time production history and, **Fig. 6.6** shows a match on the Arps side of the Fetkovich type curve. Apply the Fetkovich type curve approach to forecast performance with Fig. 6.6.

Solution. Follow these steps to obtain a solution:

1. Overlay the production data plot, Fig. 6.5 on the log (q_{Dd}) vs. log (t_{Dd}) plot of the provided Fetkovich type curve and obtain a reasonable curve fit.

Time (t) m	Rate (q_g) MMscf/D
0.1	42
6	12.7
12	4.67
18	2.28
24	1.19
30	0.74
36	0.47

Table 6.2—Redfish Well 2 production data.

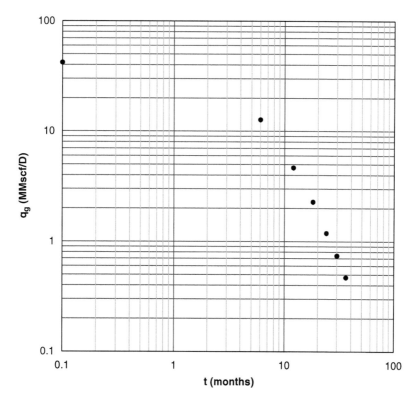

Fig. 6.5—Redfish Well 2 production data.

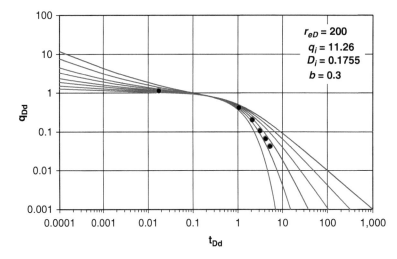

Fig. 6.6—Computer generated Fetkovich decline curve analysis match of Redfish Well 2 data. What is a particularly important characteristic of this this match?

2. Obtain *b*-exponent value and match points after curve fit.

Please note, the matched Arps (1945) decline curve stem and match points obtained from the Fetkovich type curve analysis in Fig. 6.6 are
$b = 0.3$, $t_{Dd} = 10.9$, $q_{Dd} = 0.0084$, $t = 36$ m, and $q = 0.47$ MMscf/D.

3. Apply the match points and the Fetkovich relationships to calculate (q_i) and (D_i):

$$q_i = \frac{q_z}{qDd} = \frac{0.47 \text{ MMscf}/\text{D}}{0.0084} = 56 \frac{\text{MMscf}}{\text{D}}, D_i = \frac{t_{Dd}}{t} = \frac{10.9}{36\,\text{m}} = 0.303/\text{m}.$$

The production rate at 48 months is

$$q = \frac{q_i}{(1+bD_i t)^{\frac{1}{b}}} = \frac{56 \text{ MMscf/d}}{\left[(1+(0.3)(0.303/\text{m})(48\text{ m}))\right]^{\frac{1}{0.3}}} = 0.207 \text{ MMscf/D}.$$

Cumulative production from 36 to 48 months is

$$G_p = \frac{q_i}{D_i(1-b)}\left[1-\left(\frac{q_2}{q_i}\right)^{1-b}\right] = \frac{56(30.4375 \text{ d/m})}{(0.303/\text{m})(1-0.3)}\left[1-\left(\frac{0.47}{56}\right)^{1-0.3}\right] = 7{,}753.3 \text{ MMscf}.$$

$$G_p = \frac{q_i}{D_i(1-b)}\left[1-\left(\frac{q_2}{q_i}\right)^{1-b}\right] = \frac{56(30.4375 \text{ d/m})}{(0.303/\text{m})(1-0.3)}\left[1-\left(\frac{0.207}{56}\right)^{1-0.3}\right] = 7{,}876.9 \text{ MMscf}.$$

G_p between 36 to 48 m = 7876.9 − 7753.3 = 123.6 MMscf.

Transient Example. Apply the Fetkovich (1980) type curve analysis to the M–4X well production history to estimate formation permeability and reservoir size. The general reservoir and well parameters needed for the analysis are given in **Table 6.3**, and the 160=hour. production test data are given in **Table 6.4**. **Fig. 6.7** shows the Fetkovich (1980) type curve match. The test was conducted over a 10-day period and appears to reflect mainly transient flow conditions with some transition flow regime behavior at late time.

Overlaying the production data, (**Fig. 6.8**), on the Fetkovich (1980) type curve seems to indicate that the test data remain primarily in the infinite-acting transient portion of the composite type curve. In certain cases, it can be rather difficult to determine with certainty if transition flow behavior is exhibited. In this case, as shown in Fig. 6.8, the transient flow linear decline trend is comparable in shape to that of transition flow behavior.

Well drainage volume should not be calculated from well test information alone when only transient flow data are exhibited. Sufficient transition flow regime data must be available to identify the end of infinite-acting

Reservoir rock and fluids properties

ϕ = 25%	μ_o = 2.3 cp	p_i = 3,360 psi	h = 30 ft	c_o = 1.5 × 10⁻⁵ psi⁻¹	c_w = 3 × 10⁻⁶ psi⁻¹
S_w = 27%	B_o = 1.1 RB/STB	p_{wf} = 2,660 psi		c_f = 4 × 10⁻⁶ psi⁻¹	r_w = 0.328 ft

Table 6.3—M–4X well information.

t hours	t days	q_o STB/D
6	0.25	1900
8	0.33	1800
12	0.50	1750
15	0.63	1600
18	0.75	1550
23	0.96	1500
42	1.75	1400
69	2.88	1200
110	4.58	1120
140	5.83	1080
160	6.67	1010

Table 6.4—M–4X well production data.

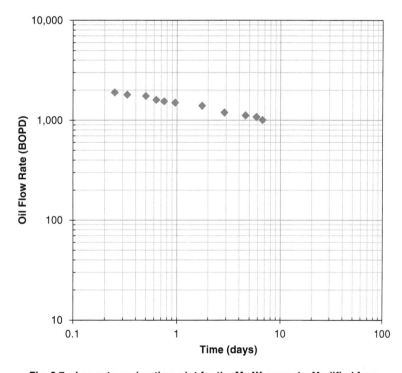

Fig. 6.7—Log rate vs. log time plot for the M–4X example. Modified from Fetkovich (1980).

transient flow before pore volume estimates can be obtained. Observation of a clear demarcation for the end of infinite-acting transient flow behavior results in greater confidence in the accuracy of the decline analysis match and corresponding analysis.

Fig. 6.8 shows the computer-generated type curve match.

Solution. Rate vs. time data and the type curve of the Fetkovich (1980) analysis. Match points are r_{eD} = 50, t_{Dd} = 0.0027, q_{Dd} = 1.780, t = 18 hours, and q = 1,550 STB/D.

1. Calculate permeability: The rate match points, r_{eD} stem value, and requisite reservoir properties are inputs.

$$k = \frac{141.2 \mu B}{h(p_i - p_{wf})}\left[\ln\left(\frac{r_e}{r_w}\right) - \frac{1}{2}\right]\left(\frac{q}{q_{Dd}}\right)_{mp} = \frac{141.2(2.3)(1.1)}{30(3360 - 2660)}\left[\ln(50) - \frac{1}{2}\right]\left(\frac{1550}{1.78}\right) = 50.5 \text{ md}.$$

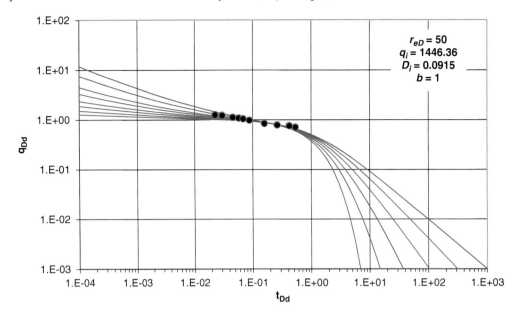

Fig. 6.8—Decline curve analysis match of M-4X test data using Fetkovich (1980) type curves method.

2. Calculate total compressibility (c_t):

$$c_t = S_o c_o + S_w c_w + S_g c_g + c_f = (0.73)(1.5 \times 10^{-5}) + (0.27)(3 \times 10^{-6}) + 0 + (4 \times 10^{-6}) = 1.58 \times 10^{-5} \text{ 1/psi}.$$

Apparent wellbore radius. The time match point values (t and t_{Dd}), matched r_{eD} stem, and the requisite reservoir properties are required to calculate the apparent wellbore radius. The units of time (t) in Eq. 6.24 are days, while the test data have been recorded in hours. A time conversion is included in the calculation to properly address the difference in the time units.

$$r_{wa} = \left\{ \frac{0.01266 k}{\phi \mu c_t \left[(r_{eD})_{mp}^2 - 1\right] \left[\ln(r_{eD})_{mp} - \frac{1}{2}\right]} \left(\frac{t}{t_{Dd}}\right)_{mp} \right\}^{0.5}.$$

$$r_{wa} = \left\{ \frac{0.01266(50.5)}{(0.25)(2.3)(1.58 \times 10^{-5})(50^2 - 1)\left[\ln(50) - \frac{1}{2}\right]} \left[\frac{18 \text{ hours}}{0.0027(24 \text{ hr/D})}\right] \right\}^{0.5} = 47.9 \text{ ft}.$$

Skin factor: $S = -\ln\left(\frac{r_{wa}}{r_w}\right) = -\ln\left(\frac{47.9}{0.328}\right) = -4.98$.

Drainage volume: $V_p = \frac{\pi(r_e^2 - r_{wa}^2)\phi h}{5.615} = \left[\frac{\overline{B_o \mu_o}}{\mu_{oi} \overline{c_t}(p_i - p_{wf})}\right]\left(\frac{t}{t_{Dd}}\right)_{mp}\left(\frac{q}{q_{Dd}}\right)_{mp}.$

$$V_p = \left[\frac{1.1}{(1.58 \times 10^{-5})(3360 - 2660)}\right]\left[\frac{18 \text{ hours}}{0.0027(24 \text{ hr/D})}\right]\left(\frac{1550}{1.78}\right) = 24{,}057{,}293 \text{ RB}.$$

The bulk volume of the reservoir connected to the well can then be determined:

$$V_b = \frac{V_p(5.615 \text{ ft}^3/\text{bbl})}{\phi(43560 \text{ ft}^2/\text{acre})} = \frac{(24{,}057{,}293)(5.615)}{(0.25)(43{,}560)} = 12{,}404 \text{ acre-ft}.$$

The reservoir drainage area of the well is obtained from the bulk volume of the reservoir drained by the well for a given net pay thickness:

$$A = \frac{V_b}{h} = \frac{12{,}404 \text{ acre-ft}}{30} = 413 \text{ acres}.$$

Utility of the Transient Side Analysis. A more in-depth study was conducted with the production test data of the M-4X well, with graphical matches made with other decline curve r_{eD} stems. A summary of four different fits that were made with the Fetkovich (1980) type curves are presented in **Table 6.5.**

Comparison of the values indicates formation permeability and skin effect can be determined with a fair degree of reliability. However, the exercise also illustrates that a quantitative estimation of the reservoir pore volume and well drainage area is more difficult to obtain when boundary-dominated flow behavior is not exhibited by the production behavior. The problem arises from similarity, (i.e., non-uniqueness), of the curvature of each of the transient r_{eD} decline curve stems.

In conclusion, we can say that results of the calculations for the infinite-acting transient portion of the Fetkovich (1980) composite decline curve are reasonable for estimating formation permeability and steady state skin effect but are much less accurate for the reservoir volumetric calculations.

The Blasingame et al. Type Curve Method

The following discussion concerns a type curve approach every bit as innovative as the Fetkovich (1980), Fetkovich and Thrasher (1979), and Fetkovich et al. (1987) methodology. Numerous investigators were involved in fleshing out the intricacies of the concept that was initiated mainly by Doublet and Blasingame (1995, 1996). However, the following discussion will be referred to collectively as the Blasingame et al. approach. Individual references might not be always mentioned in the following discussion.

Blasingame and Lee (1986) incorporated normalized rate and pressure changes ($q/\Delta p$) to develop a methodology for calculating drainage volume and Dietz (1965) shape factor. Boundary dominated conditions predominated and were applied to the concept of normalizing production rates with the pressure changes in production decline curve analyses. Palacio and Blasingame (1993) extended this work by rigorously deriving an expression for production rate normalization that was like the Blasingame and Lee (1993) expression.

$$\frac{p_i - p_{wf}}{q} = \frac{70.6\mu B}{kh} \ln\left(\frac{4A}{e^{\gamma} C_A r_{wa}^2}\right) + \frac{0.2339B}{\phi h c_t A} t_{mb} \quad \ldots \ldots \ldots \ldots \ldots \ldots \ldots \ldots \ldots \ldots \ldots \ldots \ldots \ldots \ldots \ldots (6.25)$$

Assumptions and Characteristics. The following are assumptions and characteristics:

- Bounded reservoir behavior
- Well can be located within a non-circular drainage area
- Single-phase liquid of small and constant compressibility
- $\dfrac{q}{p_i - p_{wf}}$ and $\dfrac{N_p}{p_i - p_{wf}}$

Palacio and Blasingame (1993) proved that the pressure normalized flow rate production decline relationship retraces the Arps (1945) harmonic decline curve ($b = 1$) stem on the Fetkovich (1980) type curve set. That proof provided the technical basis for developing a method for connecting transient and boundary-dominated flow regime behaviors with the harmonic decline curve. The concept also provided a basis for formulating a more

$\dfrac{r_e}{r_w}$	∞	50	100	200
k, md	45.9	50.5	49.2	57.5
A, acres	∞	227	8,462	8,462
V_b, acre-ft	∞	21,776	253,870	253,780
r_{wa}, ft	198	167.2	108.5	54.2
S	−5.5	−6.2	−5.8	−5.1

Table 6.5—Comparison of results of four graphical analyses of M-4X test data.

comprehensive type curve approach. Coalescing of all transient stems to the harmonic decline curve has been demonstrated to be valid regardless of the flow rate and flowing wellbore pressure history.

$$q_{Dd} = \frac{1}{1 + t_{mbDd}}, \quad \ldots \ldots \ldots \ldots \ldots \ldots \ldots \ldots \ldots \ldots \ldots \ldots \ldots \ldots \ldots \ldots \ldots \ldots \ldots \ldots \ldots (6.26)$$

where

$$q_{Dd} = \frac{\left(\dfrac{q}{\Delta p}\right)}{\left(\dfrac{q}{\Delta p}\right)_i} = \left(\frac{q}{p_i - p_{wf}}\right) b_{pss} \ldots \ldots \ldots \ldots \ldots \ldots \ldots \ldots \ldots \ldots \ldots \ldots \ldots \ldots (6.27)$$

$$t_{mbDd} = D_i t_p = \left(\frac{m}{b_{pss}}\right) t_p \ldots \ldots \ldots \ldots \ldots \ldots \ldots \ldots \ldots \ldots \ldots \ldots \ldots \ldots \ldots \ldots \ldots (6.28)$$

Composite Type Curve—Solutions to Reservoir Models. Fetkovich (1980) was the first to combine transient and boundary-dominated flow rate vs. time solutions into a practical, composite decline curve set. Fetkovich recognized that the rate decline solution on the transient side must converge to a boundary-dominated flow Arps (1945) exponential decline curve stem. The transient and boundary-dominated flow solutions were mathematically manipulated to force convergence along the single exponential decline solution.

Development of a type curve for a well and reservoir situation requires rate-transient solution (q_D) values for the constant pressure inner boundary condition as a function of dimensionless time (t_D) for various reservoir sizes (range of r_{eD} values). The rate-transient solutions are converted to the equivalent Fetkovich (1980) type dimensionless flow rate (q_{Dd}) and dimensionless time (t_{Dd}) function by transformations that were derived by Doublet and Blasingame (1995, 1996) and are given in Eqs. 6.29 and 6.30, respectively.

$$q_{Dd} = \left[\ln(r_{eD}) - \frac{1}{2}\right] q_D \ldots \ldots \ldots \ldots \ldots \ldots \ldots \ldots \ldots \ldots \ldots \ldots \ldots \ldots \ldots \ldots \ldots (6.29)$$

$$t_{Dd} = \left(\frac{2}{r_{eD}^2}\right) \left[\frac{1}{\ln(r_{eD}) - \dfrac{1}{2}}\right] t_D \ldots \ldots \ldots \ldots \ldots \ldots \ldots \ldots \ldots \ldots \ldots \ldots \ldots \ldots (6.30)$$

We can generate production decline curves like the Fetkovich (1980) composite decline curves, but all transient (r_{eD}) decline curve stems merge to the Arps (1945) harmonic decline curve stem in boundary-dominated flow.

Correlating Functions. McCray (1990) and others have presented graphical analysis plotting functions that include additional correlating parameters to provide more definition and clarity in the graphical decline curve analysis matching technique. Palacio and Blasingame (1993) were evidently the first to simultaneously apply multiple decline curve functions to match field data with predetermined solutions to the diffusivity equation. A set of three graphical analysis plotting functions are calculated from a solution to the diffusivity equation to provide the basis for the multiple decline type curve match. These graphical decline curve analysis functions are the

- q_{Dd}—Flow rate response function
- q_{Ddi}—Flow rate integral function
- q_{Ddid}—Flow rate integral derivative function

The integral and derivative curves are computer generated. Solutions and plotting techniques are generally driven on previously constructed computer programs, as are the accompanying Fetkovich (1980) and Poe and Poston (2010) methods. Solution procedures are shortened for this reason.

Flow Rate Function (q_{Dd}). The production decline curves for the transient behavior of an unfractured vertical well centrally located in a closed cylindrical reservoir are correlated to the harmonic decline ($b = 1$) by using the transformations presented in Eqs. 6.29 and 6.30. **Fig. 6.9** shows the transient production rate decline curves (q_{Dd}) resulting from the Doublet and Blasingame (1995, 1996) transformations.

The transient flow rate decline curves (q_{Dd} vs. t_{pDd}) presented in Fig. 6.9 are correlated as a function of the dimensionless drainage radius (r_{eD}) for the constant bottom hole flowing pressure case.

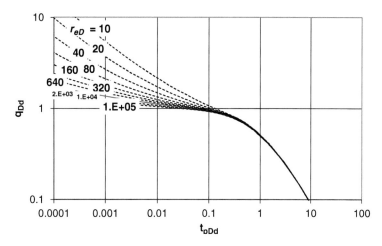

Fig. 6.9—Flow rate decline (q_{Dd}) function for unfractured vertical well. Adapted from Doublet and Blasingame (1995a).

Flow Rate Integral Function (q_{Ddi}). The flow rate integral function (q_{Ddi}) was introduced by McCray (1990). The dimensionless flow rate integral function is equivalent to the dimensionless decline time divided into normalized cumulative production, defined in Eq. 6.31. **Fig. 6.10** shows the applications of this integral transformation to the transient production rate decline curve functions for an unfractured vertical well centrally located in a closed, cylindrical reservoir.

$$q_{Ddi} = \frac{Q_{pDd}}{t_{Dd}} = \frac{1}{t_{Dd}} \int_0^{t_{Dd}} q_{Dd}(\tau) d\tau. \quad \quad \quad (6.31)$$

The flow rate integral decline curve analysis function (q_{Ddi}) trends in the same general manner as the dimensionless decline flow rate function (q_{Dd}). Unfortunately, the flow rate integral transformations produce rather symmetric, parallel-like graphical decline curve sets.

Flow Rate Integral Derivative Function (q_{Ddid}). Doublet and Blasingame (1995, 1996) introduced a third graphical decline curve analysis function to aid in differentiating production rate and flow rate integral derivative function responses. This graphical decline curve function is the negative of the derivative of the flow rate integral transformation with respect to the logarithm of the decline time. Eq. 6.32 defines the graphical decline curve analysis function.

$$q_{Ddid} = -\frac{dq_{Ddi}}{d\ln(t_{Dd})} = -t_{Dd}\frac{dq_{Ddi}}{dt_{Dd}} = q_{Ddi} - q_{Dd} \quad \quad \quad (6.32)$$

Fig. 6.11 illustrates the flow rate integral derivative function for an unfractured vertical well centrally located in a closed, cylindrical reservoir.

Note that the curvature of the flow rate integral derivative function in Fig. 6.11 can really help when selecting the proper (r_{eD}) decline curve stem to match the production decline data with.

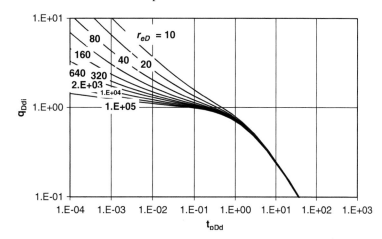

Fig. 6.10—Flow rate integral decline curve analysis function, unfractured vertical well. Adapted from Doublet and Blasingame (1995a).

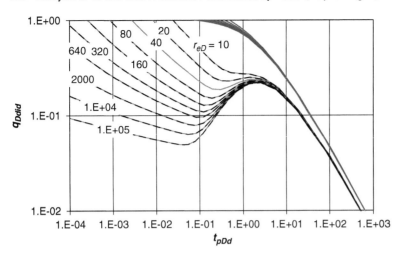

Fig. 6.11—Flow rate integral and integral derivative function, unfractured vertical well. Adapted from Doublet and Blasingame (1995a).

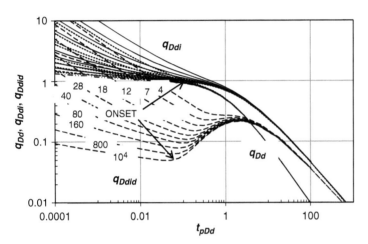

Fig. 6.12—Composite Blasingame et al. decline curve example plot of a vertical unfractured well.

Family of Curves. A composite set of decline curves incorporating the three Blasingame *et al.* decline analysis functions into a single composite decline curve analysis set has been developed from the solution to the diffusivity equation for the constant bottomhole flowing pressure case. **Fig. 6.12** shows the solution. Note that the decline flow rate (q_{Dd}) and flow rate integral function (q_{Did}) stems are visually symmetric while the flow rate integral derivative function (q_{Ddid}) provides a much clearer distinction in curvature.

Each (r_{eD}) stem is repeated for each of the curve sets. It can get confusing when trying to perform the curve analysis matching procedure manually. The flow rate, flow rate integral, and flow rate integral derivative function curves for a specific (r_{eD}) value can be easily highlighted as a set to aid when selecting the best decline curve match and (r_{eD}) stem value when analyzed with a computer-aided graphical decline curve analysis program. The decline analysis matching can be performed using all three decline curve analysis functions simultaneously to readily achieve the best match in this manner.

In summary we can say that

- The (q_{Dd} vs. t_{Dd}) plot was solved for rate-transient behavior of the well with Laplace transforms and translated the results to equivalent decline curve analysis dimensionless variable values. Solutions are presented as a function of the reservoir size (r_{eD}).
- Smoothing can be carried out by pressure drop normalization of the production rate data (if available) and by integration of the flow rate decline data with respect to decline time.
- The harmonic decline curve stem ($b = 1$) provides the right-hand side limiting value for the Arps (1945) curves, not the exponential decline curve stem ($b = 0$).
- The Blasingame et al. technique works best for smooth production decline curves. Fluctuations or outliers in the normalized field production rate decline trend can cause extreme variation in the flow rate integral derivative curve. To normalize the flow rate decline function, it must be a monotonically declining function throughout to obtain reasonable graphical analysis plotting function values.
- Matching the theoretical and field curve shapes and translating the results to the field solutions is like the process for analyzing the transient side of the Fetkovich (1980) decline curves. However, the analytically correct Blasingame and Lee (1986) and Palacio and Blasingame (1993) expression for boundary-dominated flow replaces the empirical Arps (1945) bounded reservoir models.

Reservoir rock and fluids properties						
c_t, (psi)$^{-1}$	B_o, RB/STB	μ_o, cp	ϕ, fraction V_b	Net pay, h, (ft)	S_{wi}, fraction V_p	r_w, (ft)

Table 6.6—Well and reservoir information.

Analysis Procedure. **Table 6.6** list relationships for calculating drainage area of an oil well, formation effective permeability, and skin effect from well test data.

Total compressibility (1/psi),

$$c_t = S_o c_o + S_w c_w + S_g c_g + c_f \qquad (6.33)$$

Oil in place (STB),

$$N = \frac{1}{c_t} \left(\frac{q/\Delta p}{q_{Dd}} \right)_{mp} \left(\frac{t_p}{t_{pDd}} \right)_{mp} \qquad (6.34)$$

Reservoir drainage area (acres),

$$A = \frac{NB_o}{7758 \phi h (1 - S_w)} \qquad (6.35)$$

Reservoir drainage radius (ft),

$$r_e = 117.75 \sqrt{A} \qquad (6.36)$$

Apparent wellbore radius (ft),

$$r_{wa} = \frac{r_e}{r_{eD}} \qquad (6.37)$$

Effective permeability (md),

$$k = 70.6 \frac{\mu B}{h} \ln\left(\frac{4A}{e^\gamma C_A r_{wa}^2} \right) \left[\frac{\left(\frac{q}{\Delta p}\right)}{q_{Dd}} \right]_{mp} \qquad (6.38)$$

Skin factor,

$$S = -\ln\left(\frac{r_{wa}}{r_w} \right) \qquad (6.39)$$

Matching Procedure.
Follow these steps for the matching procedure:

1. Calculate dimensionless rate integral and integral-derivative curves from field data.
2. Overlay the field data plot of ($q/\Delta p$ vs. t_p) onto the type curve set representing the solution to the correct boundary conditions.
3. Force match the (q_{Dd}) onto the harmonic stem.
4. Go to the derivative curve to obtain a secondary match.
5. Read match points and matched (r_{eD}) value. Dimensionless drainage radius (r_{eD}) is usually chosen from the integral derivative curve. Match points (t, q) and (t_{Dd}, q_{Dd}) are required to obtain the difference between dimensional field data and reference decline curve values.
6. Apply the appropriate equations and reservoir parameters to calculate oil in place and permeability values.

The same reservoir parameters obtained in a Fetkovich (1980) type curve analysis can be calculated with the new technique. Now, however there are three curves available to simultaneously match and select the best fit of the field data onto the reference decline set and the appropriate r_{eD} stem value.

Fig. 6.13 shows an example of a Blasingame et. al. plot after the three appropriate field production parameters were calculated.

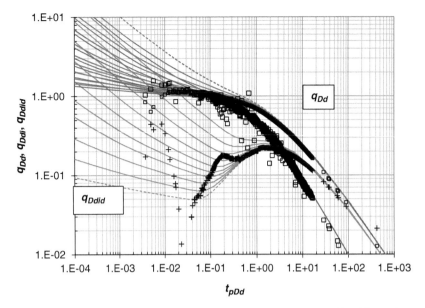

Fig. 6.13—Blasingame et al. decline curve analysis, Zuma gas well. Matching using flow rate, integral, and integral-derivative type curves on field data. Remember field curves are initially hung on the harmonic curve. Note how the derivative curve (q_{Ddid}) diverges at approximately ($r_e = 40$). Apply pertinent equations after match points are selected.

Other Blasingame et al. Type Decline Curves. Other specialized decline curve analyses have also been developed and used by Blasingame and coauthors for other well and reservoir types in addition to the decline curve solutions and analyses developed by Palacio and Blasingame (1993) and Doublet et al. (1994) for vertical wells. Doublet and Blasingame (1996) developed a decline curve solution for an infinite conductivity vertically fractured injection well in a closed, cylindrical reservoir. Production performance of a finite conductivity vertical fracture centered in a closed rectangular reservoir using the Doublet and Blasingame (1996) decline curve solution methodology has also been reported by Poston and Poe (2008). The analysis procedure for both the infinite conductivity vertical fracture solution of Doublet and Blasingame (1996) and the corresponding finite conductivity vertical fracture solution are discussed by Poston and Poe (2008).

Doublet and Blasingame (1995) also investigated water influx and waterflood performance of unfractured vertical wells using boundary flux models. Decline curves for step- and ramp-rate boundary flux models were developed for analysis of the production performance of water influx and waterflood field applications. The correlating parameters of the boundary flux models are dimensionless drainage radius, flux across the external boundary, and the time at which influx begins. The corrected transient solution, along with a summary of the analysis methodology for the Doublet and Blasingame (1995) boundary flux models, has been given by Poston and Poe (2008).

Shih and Blasingame (1995) developed a set of decline curve solutions for an infinite conductivity horizontal well using a methodology like other Blasingame et al. type decline curve solutions. The horizontal well decline curves were correlated in terms of the dimensionless drainage radius (r_{eD}), wellbore length (L_h), wellbore diameter (r_w), and completion standoff (Z_{wD}).

An empirical scaling term (ε) was added to the radial flow equation to obtain a unified boundary-dominated flow behavior by Poe and Poston (2008). The rate-transient solutions were merged into a convergent boundary-dominated flow behavior without the need for additional scaling adjustment. The transient solution for generating the rate-transient performance of an infinite conductivity horizontal well centered in a closed rectangular reservoir has been presented

Poe and Poston Type Curves

Poe and Poston (2010) applied a pressure normalization procedure to develop a family of curves merging to a single exponential decline stem. These curves and associated analyses are best suited for analyzing performance of slowly stabilizing wells completed in low permeability reservoirs exhibiting transient production behavior.

Because of its complexity, the technique requires a computer-aided analysis system in order to become practical. The matching procedure requires continuous updating of the plotting and matching graphical analysis function values. Currently, the method is not generally available for general dissemination. Only a summary of the model is presented.

Basic Adjustment Term (ε) and Outline of Method. The rate-transient behavior of a well producing during boundary-dominated flow (Eq. 6.40) provides the basis for construction of the Poe and Poston (2010) composite decline curves. The imaging term (ε) included in the radial flow equation accounts for nonradial flow effects.

$$q_D = \frac{1}{\varepsilon}\exp\left(-\frac{2\pi t_{DA}}{\varepsilon}\right) \quad\quad\quad\quad\quad\quad\quad\quad\quad\quad\quad\quad\quad\quad\quad\quad\quad\quad\quad\quad (6.40)$$

The imaging factor (ε) is a function of

- Drainage area size and shape
- Well location within the drainage area
- Type of well completion

Oil and gas rate, time, and cumulative recovery solutions for the relevant dimensionless variables are transformed to include the (ε) term to account for nonradial boundary conditions.

The Poe and Poston (2010) type curve approach predicts future performance from early transient production with a reasonable degree of certainty. The method includes plotting functions constraining the matching procedure to a single best-match position. Calculated dimensionless decline flow rate (q_{Dd}) and dimensionless pseudo production time (t_{pDd}) functions converge in a constraining behavior when plotted as a function of the dimensionless cumulative production (Q_{pD}).

Fig. 6.14 shows an example type curve. Note decreasing vertical distance between the upper and lower wings. Each type curve consists of an upper and lower wing.

- The upper wing presents the results of the calculated decline flow rate function curves obtained from analyses of the production data.
- The lower wing represents pseudo production time function curves obtained by calculating the results of the original match. A match is complete and verified when the two curves fit on a common curve set. Matching is performed on both wings of the decline curves essentially simultaneously when computer driven.

Reservoir parameters such as volume, drainage area, configuration, and permeability, as well as input values for the lower wing are calculated from the initial match points. The results of these calculations are then included into a new set of plotting functions for constructing the lower wing. Simultaneous matching of the upper and

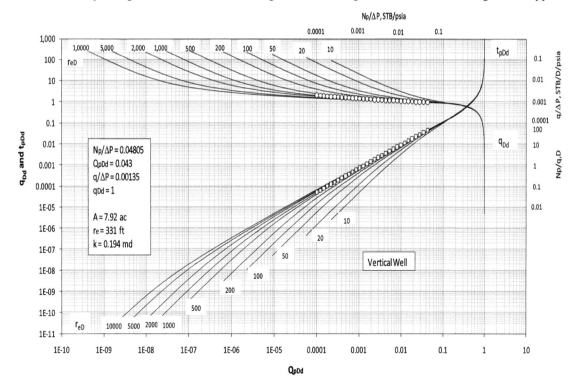

Fig. 6.14—Example vertical well decline curve match and analysis. The circles represent the resulting match of the example problem published by Poe and Poston (2010).

lower wings proves or disproves the initial interpretation. An iterative process follows if the calculated curves do not match the type curve at the proper correlating parameter values. Iterative computations of the decline curve analysis are tedious and require implementation in a computer-aided analysis to become practical.

Pressure drop normalized cumulative production and dimensionless decline cumulative production match points are combined to calculate reservoir drainage area (A) and effective permeability (k) for the well.

Single phase flow reservoir analyses are performed with only one fluid phase mobility term on the right-hand side of this expression. Applying the total mobility (md/cp) shown in Eq. 6.41 to calculate average mobility can be calculated.

$$\lambda_t = \frac{k_o}{\mu_o} + \frac{k_w}{\mu_w} + \frac{k_g}{\mu_g} \quad \text{..(6.41)}$$

Drainage area, formation effective permeability, and decline pseudo production time parameters are defined in terms of the decline analysis match position, imaging function (ε), and the dimensionless drainage area (A_D) values included in the solution procedure.

Utility of the New Plotting Method. The new method possesses several advantages over earlier methods. Those advantages are

- Permits evaluation of the drainage area, initial fluids in place, and time before the onset of boundary effects when early transient behavior predominates.
- Input and outcome values are readily calculated.
- The rate normalization procedure developed by Palacio and Blasingame (1993) simplifies the calculation procedure.
- The need for rather complicated and sensitive derivative or integral calculations and for interpretation of their relationship to the theoretical type curves is eliminated.
- Dimensionless decline pseudo production time (t_{pDd}) and dimensionless decline flow rate (q_{Dd}) are plotted as functions of dimensionless decline cumulative production (Q_{pDd}), not a time function.
- The curve set exhibits converging trends throughout the transient flow regime. These converging trends provide a unique graphical production decline analysis method.
- The method improves matches of the field performance data with the reference type curves.
- Applying early time transient production behavior provides a practical production performance interpretation technique for evaluating well and completion properties in very-low-permeability slowly stabilizing wells.
- Eliminates the need for the subjective omission of suspected production outliers in the production data to obtain a monotonic decreasing function.

Poe and Poston (2008) summarized the transient solutions used to evaluate rate-transient performance of unfractured vertical, vertically fractured, and horizontal wells in cylindrical and rectangular reservoirs. These solutions had been previously included to construct the Blasingame et al. type of decline curves.

Table 6.7 compares the various rate equations established over the course of the intellectual expansion of decline curve analysis concepts.

Field	Fetkovich	Normalized Blasingame and Lee	Radial Flow Van Everdingen and Hurst	NonRadial Poe and Poston
Time (t)	$t_{Dd} = D_i t$		$t_D = \dfrac{0.00633 kt}{\phi \mu c_t L_c^2}$	$t_{DA} = \dfrac{t_D}{A_D}$
Horner Time (t_p)			$t_{Dmb} = \dfrac{0.00633 kt_{mb}}{\phi \mu c_t L_c^2}$	$t_{pDd} = \dfrac{Q_{pDd}}{q_{Dd}}$
Rate (q_o)	$q_{Dd} = \dfrac{q_2}{q_i}$	$q_n = \dfrac{q}{p_i - p_{wf}}$	$q_D = \dfrac{141.2 q \mu B}{kh(p_i - p_{wf})}$	$q_{Dd} = \varepsilon q_D$
Cumulative (Q_p)	$Q_{Dd} = \dfrac{D_i Q_p}{q_i}$	$Q_p = \dfrac{N_p}{p_i - p_{wf}}$	$Q_{pD} = \dfrac{N_p B}{1.119 \phi c_t h L_c^2 (p_i - p_{wf})}$	

Table 6.7—Comparison of Rate – Time Equations.

Poe and Poston (2008) summarized the transient solutions used to evaluate rate-transient performance of unfractured vertical, vertically fractured, and horizontal wells in cylindrical and rectangular reservoirs. These solutions had been previously included to construct the Blasingame et al. type of decline curves. Decline curves and analysis techniques for wells that exhibit only infinite-acting transient flow conditions are included in the development. Transient decline curves and well performance analyses are presented in the following subsection for evaluating early transient performance of unfractured vertical, vertically fractured, and horizontal wells. This type of decline curve analysis has been specifically designed for use in evaluating production performance of slowly stabilizing wells, many of which are commonly completed in low-permeability reservoirs.

Completion Models. *Unfractured Vertical Well.* An unfractured vertical well located in the center of a closed circular reservoir was used as a completion model. The type curve contains (r_{eD}) stems defining reservoir size. Information obtained includes drainage area, volume, skin, permeability, and r_{wa}.

Eq. 6.42 is used for the imaging function (ε) for the unfractured vertical well case that is derived directly from the boundary-dominated flow rate-transient solution. Adjusting the imaging function (ε) is readily accomplished once the proper (r_{eD}) value has been obtained from the upper curve match.

$$\varepsilon = \ln(r_{eD}) - \frac{3}{4}. \quad\quad\quad\quad\quad\quad\quad\quad\quad\quad\quad\quad\quad\quad\quad\quad\quad\quad\quad\quad\quad\quad\quad\quad (6.42)$$

Fig. 6.15 shows the decline curves for an unfractured vertical well. The upper curves are the decline flow rate (q_{Dd}) solutions plotted as a function of the dimensionless decline cumulative production (Q_{pDd}) for a range of dimensionless drainage radius (r_{eD}) values. The lower set of curves are corresponding dimensionless decline pseudo production time (t_{pDd}) vs. dimensionless decline cumulative production (Q_{pDd}) curves. The decline curve matching procedure involves an essentially simultaneous match of both sets of decline curves.

An iterative decline curve matching and analysis procedure resolves the match for the flow rate (q) and pseudo production time functions (t_{pDd}) and the evaluation of the drainage area, formation permeability, and computed pseudo production time function values. The match procedure is complete when the selected match position of the upper flow rate decline curve and the selected drainage radius (r_{eD}) agrees with the computed pseudo production time function and matching drainage radius decline stem on the lower set of decline curves. The matching procedure continues until the selected match position and upper curve (r_{eD}) selections agree with the corresponding lower curve pseudo production time function (t_{pDd}) of the same (r_{eD}) value.

Interpreted match points are included with the analysis expressions to evaluate reservoir parameter values (A, k) and plotting function values on the lower decline curve left wing.

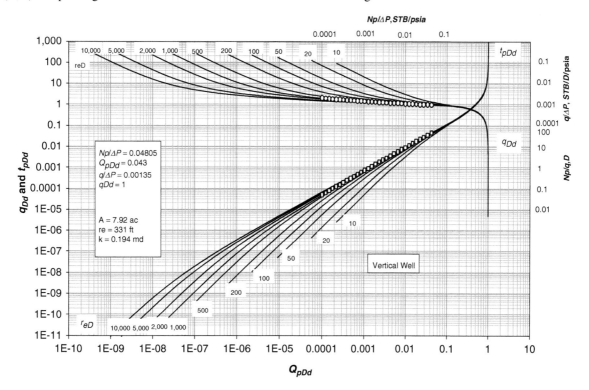

Fig. 6.15—Decline curves for vertical well in closed circular reservoir (Poe and Poston 2010).

Finite Conductivity Vertically Fractured Well. There are ten type curves, each for a different dimensionless fracture conductivity (C_{fD} = 0.5 to 500). Each decline curve stem is correlated as a function of the dimensionless drainage area (*AD*) and aspect ratio $\left(\dfrac{X_e}{Y_e}\right)$ values.

Information obtained includes drainage area, geometry, volume, skin, permeability, and r_{wa}.

England et al. (2000) found that the effective drainage area parallels the fracture plane and is essentially equal to the effective or productive "cleaned-up" fracture length. Most of the production for a vertically fractured well in a low permeability reservoir was derived from the region approximately equal to the area of the effective fracture length (X_f) or well length (L_h).

Fetkovich (1980) or Blasingame et al. decline curve analysis methods can generally be successfully applied for evaluating production performance of vertically fractured wells completed in moderate- to high-permeability formations, or even for very long-term production behavior of low-permeability reservoirs. Production performance includes transient and at least some transition or boundary-dominated flow behavior. Also, the computational effort required for the Fetkovich (1980) and Blasingame et al. analysis techniques is considerably less than that for the Poe and Poston (2010) method and is much easier to interpret, when found applicable. However, the Poe and Poston (2010) decline curve analysis technique is considerably more accurate when only transient production data are available,

Fig. 6.16 illustrates the conceptual model of a vertically fractured well centrally located in a closed, rectangular reservoir. The vertical fracture is oriented in the rectangular drainage area so that the *x*-direction is parallel to the fracture and the *y*-direction is normal to the fracture plane.

$$\text{AR} = \frac{X_e}{Y_e}. \quad\quad\quad (6.43)$$

$$A_D = \frac{A}{L_c^2} = \frac{X_e Y_e}{X_f^2} = X_{eD} Y_{eD}. \quad\quad\quad (6.44)$$

The schematic given in Fig. 6.16 indicates an increasing reservoir drainage area from that of a highly elongated rectangle with an aspect ratio of 20 to that of a square drainage area with an aspect ratio of unity. Dimensions of a square drainage area in the schematic correspond to physical dimensions of $X_e = Y_e = 2X_f$. The effective drainage area shape for a vertically fractured well completed in a low permeability reservoir will tend to be an elongated rectangle with a large aspect ratio value. Drainage area extent in the *x*-direction equals the vertical fracture length ($X_e = 2X_f$ and $X_{eD} = 2$) in each of the decline curves developed for vertical fracture transient flow analyses.

The imaging function (ε) required to merge the rate-transient solutions for the vertically fractured well into the composite decline curves are listed in Poe and Poston (2010).

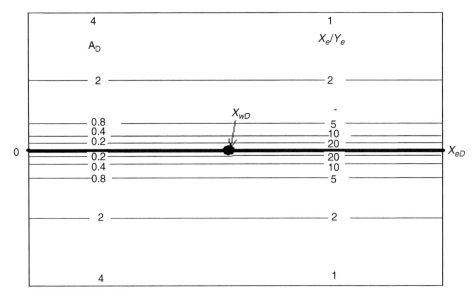

Fig. 6.16—Vertical fracture in a closed rectangular reservoir. Rectangular drainage area dimensions are correlated in terms of an aspect ratio (*AR*) and the dimensionless drainage area (*A_D*), defined in Eqs. 6.42 and 6.43, respectively.

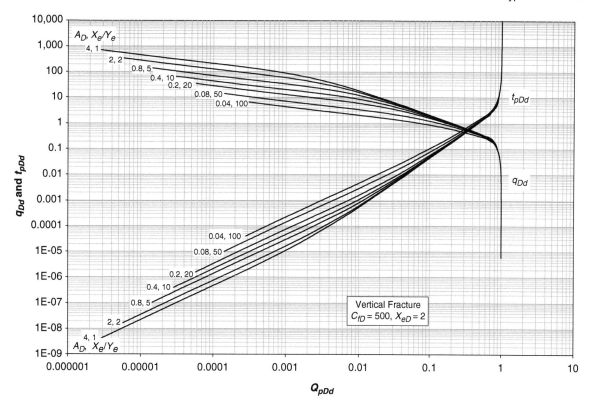

Fig. 6.17—Vertical fracture decline curves for C_{fD} = 500. The X_{eD} = 2 value was used throughout the study.

Type Curves. A set of composite decline curves was constructed for a range of dimensionless fracture conductivity values (C_{fD}) instead of a single composite production decline curve that encompasses all vertical fractures. The curvature of the line is a function of A_D and X_e/Y_e.

The curves are similar for ranges of fracture conductivity during early transient behavior (small values of dimensionless cumulative production, Q_{pDd}), which is dominated by formation linear flow before the onset of boundary effects for vertical fractures. Dimensionless fracture conductivities are usually greater than 100.

Vertical fractures with even moderate dimensional fracture conductivities ($k_f b_f$) can exhibit dimensionless conductivities (C_{fD}) in this range for the low permeability case. Fracture conductivity is sufficiently high to result in a negligible pressure drop compared with the pressure drop of fluid flow in the reservoir to the fracture during formation linear flow.

It is also noted that vertical fractures with dimensionless conductivities of approximately 5 or less ($C_{fD} \leq 5$) exhibit predominantly all bilinear transient flow at early time before the onset of boundary effects. Vertical fractures exhibiting intermediate values of dimensionless fracture conductivity exhibit bilinear flow at very early production time that is then followed by pseudolinear flow behavior before boundary effects begin to dominate the well performance. During pseudolinear flow, the well's production performance exhibits linear flow behavior, but with a finite conductivity flow component. The higher the dimensionless conductivity of the fracture, the lower the pressure drop in the fracture will be and the more the linear flow behavior will affect well performance.

The Horizontal Wellbore Case. An infinite conductivity horizontal well located in a low permeability closed rectangular reservoir was used for the horizontal wellbore case. There were seven type curves presented with each for a different dimensionless well length (L_D = 1 to 1,000). Each decline curve stem is a function of the dimensionless drainage area (A_D) and the aspect ratio $\left(\dfrac{X_e}{Y_e}\right)$. Information obtained includes drainage area, configuration, volume, skin, k, and r_{wa}.

Poston and Poe (2008) reconstructed the Shih and Blasingame (1995) decline curve solutions. The model does not require the additional scaling terms to merge the rate-transient solutions into a unified composite boundary-dominated flow model. They presented a detailed mathematical model for computing the decline curves of an infinite conductivity horizontal well in a closed cylindrical or rectangular reservoir derived from the transient solutions developed by Ozkan (1988) and Ozkan and Raghavan (1991).

Decline curves required for evaluating production performance of unfractured horizontal drainholes consist of a family of decline curve sets, just as in the case of a finite conductivity vertical well. Multiple curve sets are required because multiple parameters affect well performance. Dimensionless well drainage radius (r_e), (or x- and y-direction drainage areal extent), wellbore length (L_h), wellbore radius (r_w), standoff (Z_w), and permeability anisotropy affect the transient performance for an infinite conductivity horizontal wellbore.

Assume that the horizontal wellbore possesses infinite conductivity and a rectangular drainage pattern depleting from a low permeability matrix. The well and reservoir configuration is like Fig. 6.16 for a vertically fractured well.

Model development is like the vertical fracture case, except that the extended wellbore reach must be included in the geometry of the system, with the wellbore position in relation to the upper and lower bed boundaries.

Dimensionless parameters are required to characterize performance of the horizontal well in the reference decline curve set. Assume that the effective drainage extent parallels the horizontal wellbore (x-direction) and is approximately equal to the effective length of the horizontal drainhole (L_h). The well and reservoir schematic shown in Fig. 6.16 depicts the well and reservoir configuration.

Studies have shown infinite conductivity horizontal well decline curves can be reasonably well characterized with only two parameters when the range of applicability is limited to the early transient behavior of a horizontal well completed in a low permeability reservoir.

Imaging function relationships used to construct the type curves for an infinite conductivity horizontal well in terms of the drainage area (A), aspect ratio (AR), dimensionless drainage area (A_D), and wellbore length (L_H) are shown in Poe and Poston (2008). Computed imaging function values for a range of dimensionless drainage area, aspect ratio, and wellbore lengths are summarized in that paper.

A comparison of the reference decline curve solutions for an infinite conductivity horizontal well in a closed square with a dimensionless wellbore radius (r_{wDz}) equal to 0.002 is presented for a range of dimensionless wellbore lengths in **Fig. 6.18.**

In summary we can say that the Poe and Poston decline curve analysis method lends itself to the analysis of wells undergoing transient depletion conditions.

Problems

Example Problem 6.1. Apply the Fetkovich type curve analysis technique to analyze the Oklahoma Alicia #1 production history shown on the plots below:

- Semilog rate vs. time plot in **Fig. P6.1.1**
- Log rate vs. log time plot in **Fig. P6.1.2**

Learning Objective. Expand knowledge of reservoir and well character by interpretation of performance with Fetkovich type curve.
What side of the Fetkovich type curve provides the best fit?

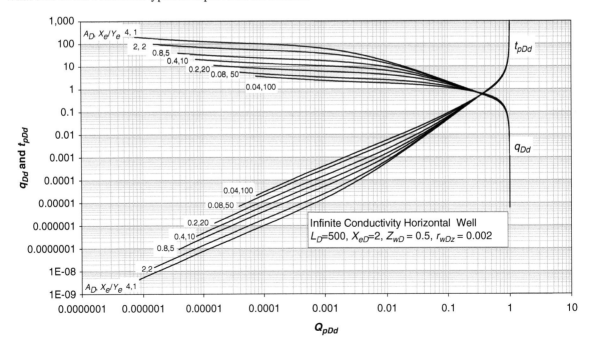

Fig. 6.18—Decline curves for an infinite conductivity horizontal well, L_D = 500 (Poe and Poston 2010).

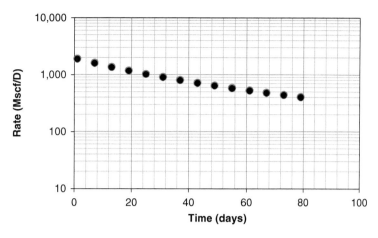

Fig. P6.1.1—The semilog rate vs. time plot of Alicia #1 well. What boundary situation does the shape of the performance curve seem to indicate?

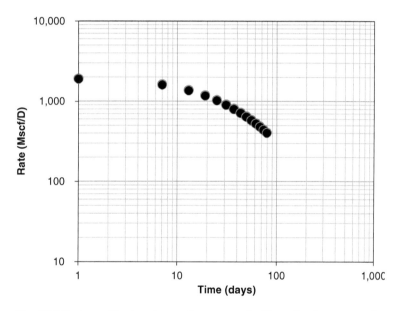

Fig. P6.1.2—Logarithmic rate vs. time curve to fit on the type curve for Oklahoma Alicia #1. Overlay Fig. P6.1.2 onto the transparency to obtain a curve fit, the *b*-exponent value, and match points.

Example Problem 6.2. The semilog rate time plot for the Drew Buzz well is shown in **Fig. P6.2.1** while the logarithmic rate vs. time plot to overlay on the Fetkovich type curve is shown **Fig. P6.2.2**.

Learning Objective. Expand knowledge of reservoir and well character by interpretation of performance with Fetkovich type curve.

Determine the appropriate *b*-exponent value and match points. What do you think about the quality of the curve fit?

Example Problem 6.3. Analyze **Figs. P6.3.1 through P6.3.5** to determine the productive character of the Golden Zuma well.

Learning Objective. Expand knowledge of well character by interpretation of performance with different variables.

Table P6.3.1 lists reservoir and production parameters and **Fig. P6.3.1** displays the performance history. Analyze each of the plots provided in order to determine reservoir and production properties.

The performance plot (Fig. P6.3.1) shows evidence of transient and boundary dominated flow and marginal production fluctuation.

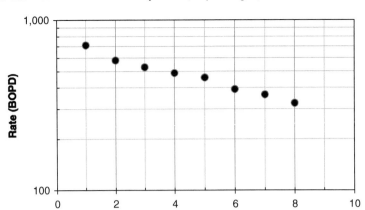

Fig. P6.2.1—Semilog rate vs. time plot of Drew Buzz well.

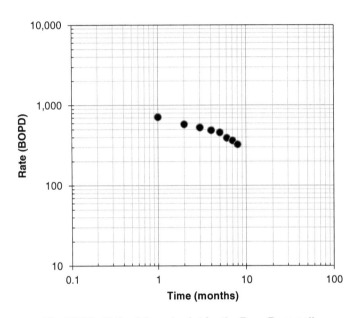

Fig. P6.2.2—Fetkovich scale plot for the Drew Buzz well.

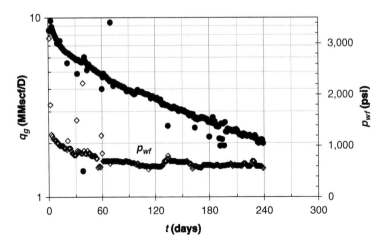

Fig. P6.3.1—Performance plot.

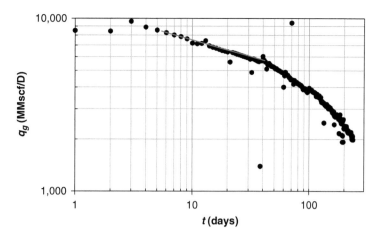

Fig. P6.3.2—The ln$_q$ vs. ln t plot. The concave downward line at late time indicates the producing interval is probably not homogeneous. Boundary influence after 50 days or after 150 days?

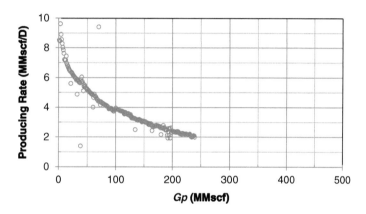

Fig. P6.3.3—Rate vs. cumulative - recovery plot for the Zuma well. How satisfied are you with the straight line fit? Seems to indicate EUR = 470 MMscf?

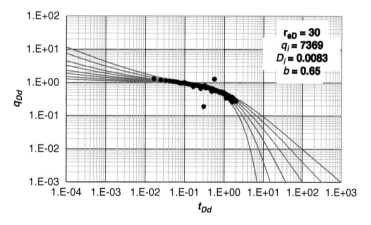

Fig. P6.3.4—The Fetkovich decline curve match for the Zuma gas well. The plot indicates both transient and boundary dominated flow occur.

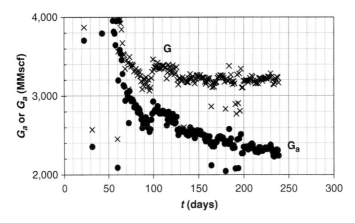

Fig. P6.3.5—Original gas in place plot for the Zuma well.

h = 54 ft	c_t = 3.2 x 10^{-4} psi^{-1}	r_w = 0.328 ft	γ_g = 0.68
ϕ = 27%	$1/B_{gi}$ = 245 scf/rcf	p_i = 3,670 psi	μ_g = 0.017 cp
z = 0.86	B_{gi} = 0.0044 rcf/scf	T = 196°F	q_{EL} = 200 Mscf/D

Table P6.3.1—Golden Zuma Gas Well and Reservoir Information.

Both transient and boundary dominated conditions existed. Essentially constant (p_{wf}) conditions exist for most of the well life.

- Estimate (q_i) in order to apply quadratic equation.
- Estimate D_{exp}.
- Estimate EUR.
- What can you say about the quality of the production data?

Match points are b = 0.65, t = 10 days, t_{Dd} = 0.083, q = 1,000 Mscf/D, q_{Dd} = 0.1357.

Analyze Arps Side. Calculate cumulative production over the production time at which the production rate decreases from 2,000 Mscf/D to the economic limit of 200 Mscf/D.

- Calculate expected production. ΔG_p =
- What is the time interval over which the production decline occurs?
- Estimate EUR.

Analyze Transient Side. Calculate the formation permeability. Match points, r_{eD} = 30 value, and requisite reservoir properties are entered.

$$k = \frac{1422 T \mu Z}{h\left(p_i^2 - p_{wf}^2\right)} \left[\ln\left(\frac{r_e}{r_{wa}}\right) - \frac{1}{2}\right]\left[\frac{q}{q_{Dd}}\right]_{mp}.$$

- Calculate the apparent wellbore radius:

$$r_{wa} = \left\{\frac{0.01266 k}{\phi \mu c_t \left[(r_{eD})_{mp}^2 - 1\right]\left[\ln(r_{eD})_{mp} - \frac{1}{2}\right]}\left(\frac{t}{t_{Dd}}\right)_{mp}\right\}^{0.5}$$

- Evaluate Skin Effect:

- Calculate the reservoir drainage pore volume. Note that the gas viscosity terms are assumed approximately equal and cancel out.

$$V_p = \frac{\pi(r_e^2 - r_{wa}^2)\phi h}{5.615} = \left[\frac{56.557 T \overline{\mu_g} \overline{Z}}{\mu_{gi} \overline{c_t}(p_i^2 - p_{wf}^2)}\right]\left(\frac{t}{t_{Dd}}\right)_{mp}\left(\frac{q}{q_{Dd}}\right)_{mp}$$

- The bulk volume of the reservoir drained by the well:
- The corresponding drainage area of the well:

Quadratic Equation. Input values are $q_i = 6.0$ MMscf/D, $p_i = 3{,}670$ psi, and $z_i = 0.94$. Apply the quadratic equation concept to calculate the original gas in place for the Zuma well.

- Recall: $G_a = \dfrac{1}{2}\dfrac{\left[2\eta \pm 2(\eta^2 + \eta q_g - \eta q_{gi})^{0.5}\right]G_p}{q_{gi} - q_g}$, in MMscf.

- Let $A = 2(\eta^2 + \eta q_g - \eta q_{gi})^{0.5}$, in MMscf/D. ($G_a$) can be shortened to $G_a = \dfrac{1}{2}\left[\dfrac{(2\eta \pm A)G_p}{q_{gi} - q_g}\right]$.

- The (p/z) ratio relationship is $\left(\dfrac{p_{wf}/z_{wf}}{p_i/z_i}\right) = \left(\dfrac{p_{wf}/z_{wf}}{3{,}670/0.88}\right) = \left(\dfrac{p_{wf}/z_{wf}}{4{,}267}\right)$.

- The (η) term when $p_{wf} = 1{,}220$ psi is $\eta = \dfrac{q_{gi}}{\left[1 - \left(\dfrac{p_{wf}/z_{wf}}{p_i/z_i}\right)^2\right]} = \dfrac{6.0 \text{ MMscf/D}}{\left[1 - \left(\dfrac{1220/0.94}{4{,}267}\right)^2\right]}, = 6.61$ MMscf/D.

Example Problem 6.4. Was the workover of the Kentucky well successful?

Learning Objective. Apply the type curve approach to show how reinitializing a well history can help when analyzing well performance.

The producing section of a Kentucky well consists of a multilayered interval. **Fig. 6.4.1** shows the semilog rate vs. time plot for the well. A normal production decline existed until Month 11 when the well was stimulated. to increase the producing rate. Economic limit is 180` BOPM.

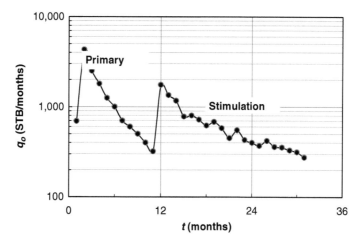

Fig. P6.4.1—Production history. The stimulation treatment at Month 11 caused a dramatic change in the producing rate.

Performance history can be divided into two segments: primary and remedial. Did stimulation increase reserves?
Analysis Procedure. The production curve was reinitialized for the remedial phase and **Figs. P6.4.2 and P6.4.3** show these segments on log - log plots equivalent to Fetkovich scale.

- Calculate remaining production for the primary depletion match.
- Calculate expected production for the remedial phase.
- Compare the two decline histories to determine the effect of the workover.

A match of field and theoretical curves was made and the match points were $b = 0.4$, $t_{Dd} = 1.0$, $q_{Dd} = 0.1$, $t = 1.85$ months, $q = 900$ BOPM.

- Calculate (q_i and D_i) in order to apply the Arps hyperbolic equations.
- Calculate cumulative production for the primary stem.

Match points are $b = 1.0$, $t_{Dd} = 1.0$, $q_{Dd} = 0.1$, $t = 1.2$ months, $q = 300$ BOPM.

- Calculate remaining time to reach economic limit. By inspection, $t = 35$ months when $q = 180$ BOPM.
- Calculate cumulative production to the economic limit.

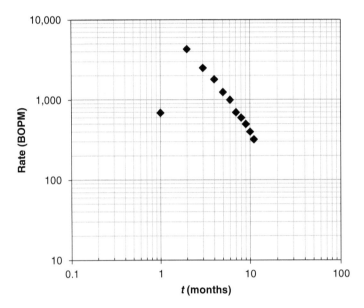

Fig. P6.4.2—The log-log plot for primary depletion phase of Kentucky well. By inspection, production would have lasted for 15 months trending down to the economic limit.

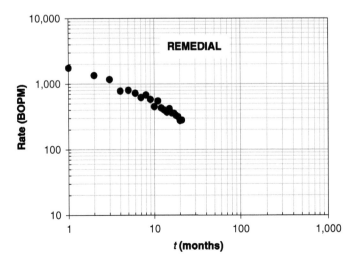

Fig. P6.4.3—Log-log plot of the "remedial" depletion stem.

Chapter 7
Two Phase Flow

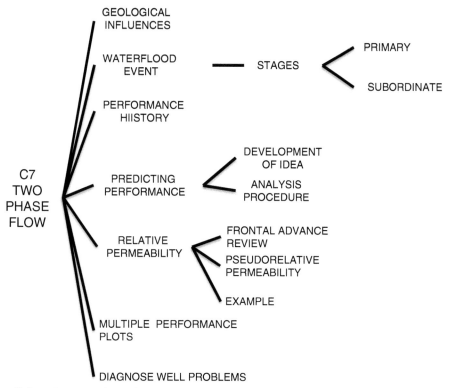

Declining oil production might be caused by

1. Declining producing rate and reservoir pressures as a function of time. Normal decline curve analysis techniques are often applied to predict future performance.
Or by
2. Increasing secondary fluid volumes that enter the well stream. Introduction of an additional phase to the flow stream reduces hydrocarbon relative permeability and available cross-sectional area in the tubing open to the oil phase, hence reducing oil production.

Waterflooding can yield incremental production when primary recovery is not very efficient. Injecting water to augment reservoir pressure and push oil toward the pressure sink of a completion is the common method for increasing recovery from pressure-depleting reservoirs. Water injected into the reservoir or into the aquifer downdip of the reservoir increases reservoir energy by replacing fluids withdrawn in primary recovery and repressurizing the system.

However, constant vigilance is required to monitor waterflood operations because of the organic nature of the process. Interaction between producing- and injection- well patterns, physical plant capacity, and complex reservoir geology encourages a constantly changing atmosphere. Maximizing waterflood efficiency is predicated on the assumption of constant surveillance. Compromising on this assumption can lead to bad results.

Recognizing two phase flow relationships such as gas/oil ratio (GOR), water/oil ratio (WOR), or fraction of water flowing (f_w) as a function of time is the primary method for identifying changes to the flow system. The following discussion introduces the concept of applying multiple performance plots to develop an understanding of reservoir and well behavior when more than a single fluid is flowing. These concepts are divided into five categories.

1. Study production plots to review the operational history as an aid when forecasting future performance and estimating remaining potential.
2. Develop relative permeability curves from performance data to aid in reservoir analysis and simulation studies.
3. Apply the Arps equations and segment concepts to estimate reserves and predict future performance.
4. Apply multiple performance plots to aid in predicting performance.
5. Apply derivative curves to diagnose well problems.

Geological Influences

The presence of single- or multiple-pay sands within a producing zone can materially affect oil and water production history. Lithology has an important influence on the efficiency of water injection in a particular reservoir. Reservoir lithology and rock properties such as porosity, permeability, clay content, and net thickness will affect waterflood ability and success. Completing in the oil portion of homogeneous sand (Case A) in **Fig. 7.1** provides no impediment to water migrating upward from the water leg. The oil/water contact (OWC) will gradually rise until viscous forces cause water to cone upward in opposition to gravity.

Upward movement of water for Case B is hindered by the presence of a shale barrier. Sand members deplete as a function of the encroaching water front in that particular layer, not as a whole. So, variation in reservoir properties either vertically or horizontally as a function of space (heterogeneity) has major impact on waterflood projects.

Fig. 7.2 shows two types of waterdrive displacement mechanisms. Recovery efficiencies are profoundly different.

Edgewater drive is a function of the reservoir dip, oil viscosity, and permeability. Water sweeps oil and gas ahead of the interface to reduce oil and/or gas saturation as a function of sweep efficiency. Edgewater drive is the more desired of the waterdrive mechanisms.

Bottomwater is a function of the interplay between gravity and viscous forces. Water coning upward in opposition to gravity reduces the area sweep efficiency by gradually increasing water production, which consequently reduces oil production until the economic limit is reached.

Shale layers of minor thickness can profoundly affect well performance history.

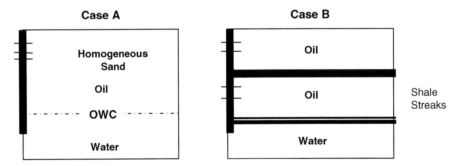

Fig. 7.1—Effect of shale layers on production from waterdrive reservoirs. The lack of an impediment in Case A allows water coning to eventually occur, which in turn causes the well to water out much sooner than in Case B where shale layers inhibit initiating a water cone.

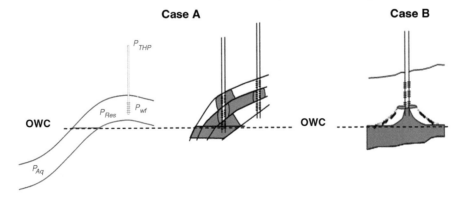

Fig. 7.2—Case A: Edgewater—encroaching water sweeps the lighter density gas or oil ahead of the interface (OWC). Case B: Bottomwater—water coning causes an increasing WOR. Excess water production gradually causes operating costs to exceed income.

The Waterflood Event

Fig. 7.3 provides an idealized example of the flood-front pattern of a five-spot waterflood. Water displacing oil moves outward in a two-step process during the waterflood process.

- In Step 1, initially the water/oil interface is displaced outward in a piston-like manner. Basically, a barrel of injected water displaces a barrel of oil until the breakthrough event occurs. In effect, we have oil at $S_w = S_{wi}$ ahead of the front and $S_w = 1 - S_{or}$ behind the front. Only oil is produced until breakthrough.
- In Step 2, oil and water phases flow simultaneously through the porous media after breakthough. The growth rate of the swept area decreases because of the reduction of injected water volume displacing oil.

Cobb and Marek (1997) coupled material balance techniques with Buckley and Leverett (1942) theory to predict performance in mature waterfloods. **Fig.7.4** illustrates the envisioned model, which is similar to the Stiles (1949) concept of flooding out a layered reservoir. Water advancement in individual layers is a function of the water- to oil-mobility ratio and permeability contrast between layers.

$$M = \frac{\text{Mobility of Water}}{\text{Mobility of Oil}}$$

$$= \frac{\dfrac{k\,k_{rw}}{\mu_w}}{\dfrac{k\,k_{ro}}{\mu_o}} = \frac{k_{rw} * \mu_o}{k_{ro} * \mu_w} \quad\quad\quad\quad\quad\quad\quad\quad\quad\quad\quad\quad\quad\quad\quad\quad\quad\quad\quad\quad\quad (7.1)$$

When $M = 1$ is neutral, water and oil move equally well. When $M < 1$ is favorable, oil will move easier than water. When $M > 1$ is unfavorable, water will move easier than oil.

Intrawell behavior in a layered reservoir can be affected by a number of factors. For instance,

- Variation of the permeability distribution between the water injection wells and producing well(s)—i.e., well pattern.
- Degree of crossflow

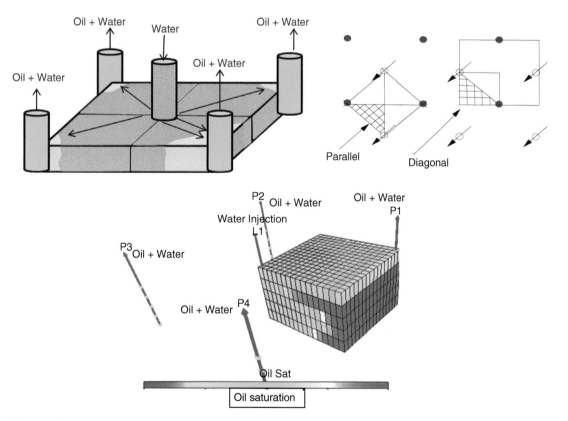

Fig.7.3—Growth of the water/oil contact in a five-spot waterflood. The interface is not uniform because of variations in rock character. Also shown, 3D visualization of a diagonal grid.

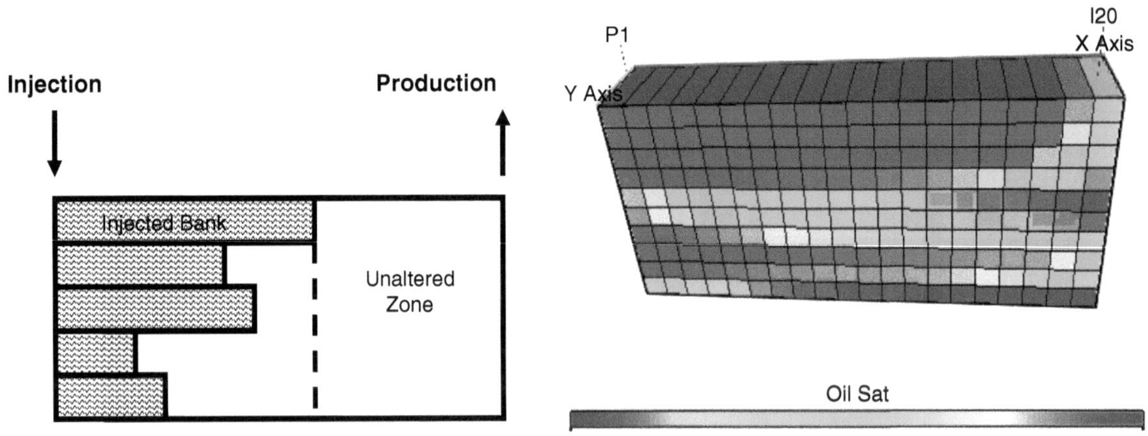

Fig. 7.4—A stratified reservoir showing flood front before fill-up. Note, breakthrough of the oil bank has occurred in only three of ten layers of the stratified reservoir simulation in the figure on the right. It shows a degrading from red color at S_{oi} to, different grades of oranges, yellows, greens and intense blue ($S_w = 1-S_{or}$).

- Variation of rock properties
- Relative injection and production pattern in the field

There are three different forces operating during a waterflood: viscous force, gravity force, and capillary force.

- Viscous Force
    - Because of pressure gradients imposed
    - Mainly controls horizontal fluid movement
    - Related to viscosity
- Gravity Force
    - Because of fluid density difference
    - Controls gravity segregation and vertical movement

There is a viscous/gravity number that

- Serves as an indicator of importance of gravity force in the displacement process
- Ignores capillary forces

$$N_{gv} = \frac{K_v \, k_{rw} \, \Delta\rho \, g \, \cos(\alpha) \, A}{887.2 q \mu_w} \frac{L}{h}. \quad\quad\quad (7.2)$$

Field units are used in Eq. 7.2. α is the dip angle from horizontal. The higher the dip, the lower the viscous/gravity number (N_{gv}).

The viscous dominates when $N_{gv} < 0.1$. Gravity dominates when $N_{gv} > 10$. The flow is "transitional" when $0.1 < N_{gv} < 10$. When the N_{gv} number is intermediate, the flow is "transitional" and shows characteristics between the gravity dominant and the viscous dominant flow.

- Capillary force
    - Active at interface between immiscible fluids
    - Generally weak force in fluid displacement
    - Has significant effect in establishing residual saturations

There is a capillary/viscous number

$$N_{cv} = \frac{K_v \, A \, L \, \Delta(P_c)}{887.2 q \mu \, h^2} \quad\quad\quad (7.3)$$

The capillary/viscous number is an indicator of importance of capillary force in the displacement process. It also ignores gravity. Better floods occur in high N_{cv}. For N_{cv} greater than 10, waterflood would show the following characteristics:

- A uniform flood front
- Sustained peak oil rate
- Late water breakthrough
- Rapid watering out of wells after breakthrough
- Small post-breakthrough oil recovery

The well-known Stiles (1949), Buckley and Leverett (1942) and Cobb and Marek (1997) techniques require simplifying assumptions and idealizing the fractional flow curve, which usually does not mirror actual field character. Pseudorelative permeability curves are developed for reservoir simulation.

Water Breakthrough and Effects on Production. Fig. 7.5 shows a (WOR vs. N_p) plot reflecting the water breakthrough events for a two-layer system; MBO means 1,000 BO. The encroaching water front moves more easily through the highest-permeability layers. Initially, water breakthrough to a producing well will occur in the most conductive layer while lower-permeability layers are still displacing oil toward the wellbore.

The effects of waterflooding heterogeneous reservoirs are often monitored by replacing the cumulative water injection term (W_i) on the x-axis with cumulative recovery (N_p). Application of either method produces curves of similar character.

The straight-line relationship of the semi-log plot is readily extrapolated into the future. However, the logarithmic scale compresses the WOR values. An alternative and more-expressive plotting method is the WOR vs. cumulative production plot.

Example. The production history of a dual layer waterflood of the Cypress (high permeability) and Benoist (much lower permeability) sands located in a 700-acre unit is shown in **Fig. 7.6** with a plot of log WOR vs. N_p. This figure is adapted from Timmerman (1971).

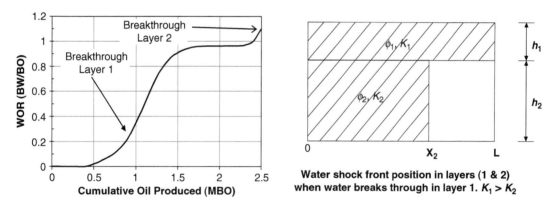

Fig. 7.5—A field (WOR) curve from a layered reservoir may exhibit a series of step-like produced water volume increases as a function of time. Notice, how cumulative oil production (N_p) has been inserted for the change in water saturation on the abscissa.

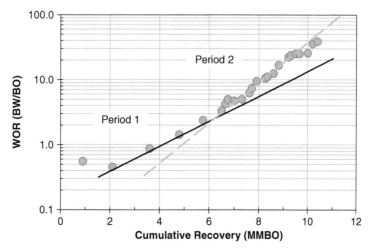

Fig. 7.6—WOR history of the Cypress and Benoist sands of differing permeability. Adapted from Timmerman (1971).

Two straight line approximations of the performance curve are present. The first straight line represents Period 1 and reflects the combined production from both sands. The second straight line representing Period 2 occurs after the Benoist sand flooded out and the lower-permeability Cypress sand provides the majority of the produced oil. The change in the slope of the line after 6.7 million BO (MMBO) of cumulative production indicates that the depletion process changed from flooding two zones to a single zone. The project becomes less effective.

Ershaghi and Omorigi (1954) and Ershaghi and Abdassah (1984) developed a linearization technique for generating a composite relative permeability curve from production records. They observed that a straight line in a water cut vs. recovery curve is the product of reservoir configuration, rock heterogeneity, and displacement efficiency. Therefore, the shape of performance curve represents the combined effects of layering, permeability variation, rock and fluid properties, well layout, and surface operations.

Reorganization and application of oil, water, and gas production records into relative permeability changes would replace the idealized laboratory fractional flow curve with a field-derived fractional flow curve. The technique also divides reservoir or well histories into representative producing intervals, with each segment reflecting essentially constant and continuous operating conditions.

Assumptions. The assumptions are

- Applicable for mature field when an invading fluid displaces an in-situ fluid (e.g., in an ongoing waterflood) where $p_{res} < p_b$ and $f_w \geq 50\%$.
- Recovery efficiency is a function of the fractional flow (f_w) vs. N_p curve developed from the field WOR plot.

Depletion Stages. There are two depletion stages. To better understand these stages, divide a flood-out history into a primary and a subordinate stage. The primary stage occurs when injection commences with streamlines advancing toward the major pressure sinks through the highest-permeability rocks. Therefore, we find increasing sweep efficiencies because of the positive displacement character measurably expanding the swept volume.

The subordinate stage occurs when the gross outlines of the flow system have been attained. Continued oil production is a function of depleting oil from the upper end of the fractional flow curve during the subordinate stage. Generally, $f_w > 50\%$ is a result of water cycling effects.

Measurable changes such as infill drilling, a significant workover program, or changes in the injection-/ production-well geometry affect the shape of the production derived fractional flow curve. These changes influence future reserves estimation and can be good or bad.

If appropriate, divide field history into a series of producing stages or segments, with each one reflecting an essentially constant well count and injection-/production-well locations. Relationships between the two stages are a function of

- Permeability's and permeability distribution in the *x*-, *y*-, and *z*-planes.
- Fluid properties.
- Well distribution and locations.
- Water injection and production volumes.
- Water cut being greater than 50%.
- Voidage to replacement (injection) ratio close to unity.
- Well count and operating conditions remain relatively constant.
- Subsurface and field operating architecture remain the same.

Example Waterflood Review. An example of waterflood review is shown by using analysis of the Provost Lloydminster "O" Pool discussed by Baker et al. (2003). In Chapter 2, we showed the same example and discussed the importance of identifying different production segments. In this section, performance plots that portray the major aspects of the field history are discussed and analyzed using the water injection effects that commenced in January 1977. General characteristics, include OOIP = 10.1 m^3, recovery = 50%, API gravity = 23.8 °API, and thickness = 10.7 m.

Performance plots to interpret field performance are presented in the following figures.

- **Figs. 7.7 through 7.9** present performance history, relating production parameters to history.
- **Fig. 7.8 to 7.10** presents segments of the field history.
- **Fig. 7.9 to 7.11** presents curve fitting of the depletion side.
- **Fig. 7.10 to 7.12** presents recovery factor (RF) vs. hydrocarbon pore volume injected (HCPVI).
- **Fig. 7.11 to 7.13** presents WOR vs. cumulative oil recovery which, projects future water production.

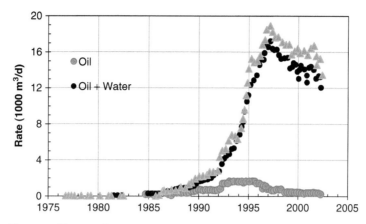

Fig. 7.7—Oil, total liquid producing rate, and water injection history for the Provost Lloydminster "O" Pool. It is apparent that water cycling is of major concern for this waterflood because only a small fraction of produced fluids is oil. Adapted from Baker et al. (2003).

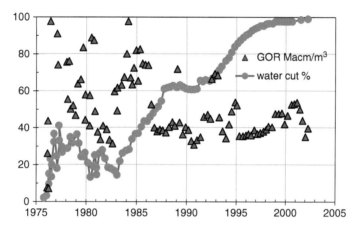

Fig. 7.8—GOR and water-cut performance history from Provost Lloydminster "O" Pool. Adapted from Baker et al. (2003).

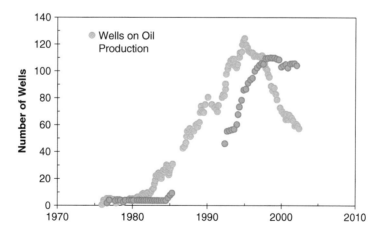

Fig. 7.9—Well history water handling costs are the main driver for an economic waterflood. A large number of water cycling oil wells have been shut in due to lack of income production. Producing wells displaying an excessive WOR are actively being shut in when uneconomic. Provost Lloydminster "O" Pool. Adapted from Baker et al. (2003).

Performance History—Provost Field

Figs. 7.7 through 7.9 provide an analysis of the historical producing history for the Provost Field.

Similarity of the water injection and oil and water volumes indicate little actual volumetric displacement within the field after general water breakthrough.

The Fig. 7.8 curve set shows the water cut and GOR history. As would be expected, early GOR was >200 but it rapidly declined to 120 as gas evolution was halted by the water injection project.

Water cut increased over the life of the project and attained the 50% value at approximately 11 years after project commencement. Currently water cut is approximately 97%.

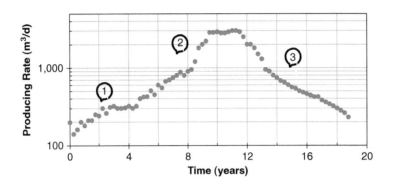

Fig. 7.10—Segments of the Provost waterflood. The field has entered the last stages of depletion. Adapted from Baker et al. (2003).

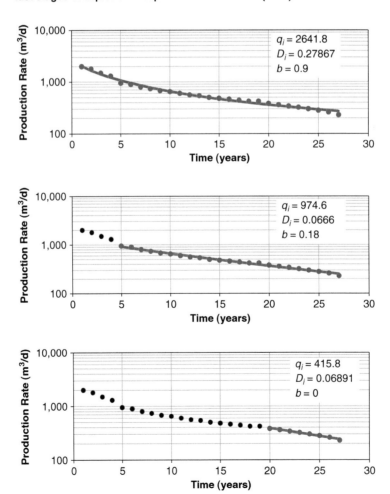

Fig. 7.11—Comparison of the Arps curve fits.

Segments of the Waterflood. Fig. 7.10 shows the performance divided into three segments.

1. Primary depletion and initiation of the waterflood at Year 4.5.
2. Significant oil displacement occurred with major flood out of the more permeable layer, which continued until Year 11.
3. Mature waterflood, with water cut > 50%. Water production through the more permeable streaks became the major contributor. Remaining movable oil is generally lodged in the smaller pore spaces where capillary forces predominate. Depletion enters a more gradual phase because areal and vertical sweep efficiency have become essentially static.

Comparison of Predictions. As discussed in Chapter 2, Baker et al. (2003) showed that an equally good fit of the Arps equation could be achieved with a b-exponent = 1.0, 0.36, or 0 according to the start of the fit.

Fig. 7.11 compares curve fits for *b*-exponent = 0.9, 0.18, 0.0 and shows the length of the approximating curve. **Table 7.1** compares the results.

	HISTORY		
1 → 27 (years)	$b = 0.93$	$q_i = 2{,}642$ m³/d	$D_i = 0.28$/yr
5 → 26 (years)	$b = 0.18$	$q_i = 975$ m³/d	$D_i = 0.061$/yr
20 → 27 (years)	$b = 0.0$	$q_i = 416$ m³/d	$D_i = 0.069$/yr

Table 7.1—Comparison of Predictions.

Which curve fit appears to be the best? What is happening with the last two data points.

Fig. 7.12 below shows that the waterflood has been managed efficiently given the recovery obtained of approximately 40% of the original oil in place after 2 hydrocarbon pore volumes had been injected. It can be observed also, that ultimate recovery can be estimated in 50% of the original oil in place by injecting a total of 5.5 hydrocarbon pore volumes.

Water handling costs increase as a function of time in a waterflood. Water cleanup and injection costs can become of increasing importance, and eventually producing costs become greater than income and the project is shut in. **Figs 7.7 and 7.13** shows that the latest water rate is rapidly increasing. In fact, Fig. 7.9 indicates that a number of wells have been abandoned. The cause is probably excessive water cycling.

Summary of Analysis. In summary,

- The waterflood has produced most of the reserves.
- Water handling costs are of paramount importance, and continued operation is tied to net income.
- Close monitoring of field wells and shutting them in when oil production declines below the economic limit optimize field expense and maintain field viability.

Estimating Reserves and Predicting Performance

Arps's equations are applied to develop a method for estimating reserves and forecasting oil and water producing rates from rate vs. time data. The exercise forecasts future oil producing rates and expected water handling costs into a cash flow stream.

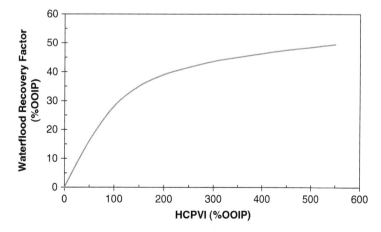

Fig. 7.12—Slope of line abruptly flattened after significant water breakthrough. The waterflood appears to be very efficient because nearly 40% recovery was achieved after approximately 2 HCPVs of water injection. Adapted from Baker et al. (2003).

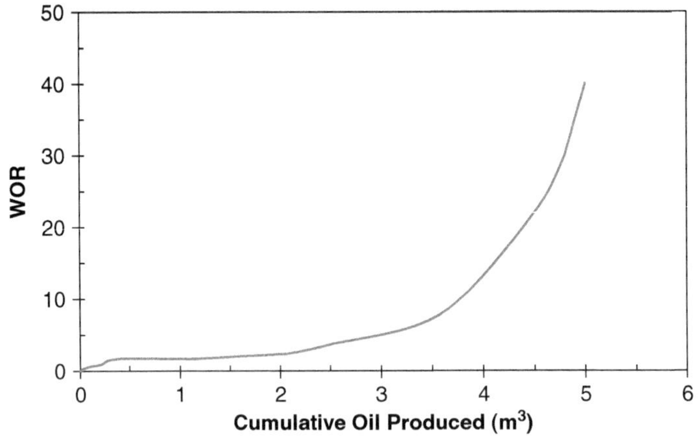

Fig. 7.13—The plot accentuates increase of water handling costs as a function of cumulative production and appears linear at late time. The WOR remained moderate until breakthrough was achieved. Is water cycling evident? Adapted from Baker et al. (2003). Equating oil price to the cost of production and water handling became most important after 4.8 m³ STBO had been produced.

Exponential equations (Eqs. 1.6 and 1.11) are applied in this case. Hyperbolic equations work equally well. The rate equation is

$$q_2 = q_1 \exp(-Dt). \quad\quad\quad (7.4)$$

The cumulative production expression is

$$N_p = \frac{q_1 - q_2}{D}. \quad\quad\quad (7.5)$$

Assumptions and Fundamentals. The assumptions and fundamentals are

- Rate vs. time and WOR plots can be divided into one or more segments.
- Apply Arps exponential or hyperbolic equations to predict future performance. One should be cautious in applying high b-exponent values.
- Field plant (producing/injection arrangements) does not change appreciably.
- Apply the WOR vs. N_p plot to predict future water production.

Solution Procedure. Begin the process by setting up a table (**Table 7.2**) outlining the characteristics of the last known producing segment. Include estimates of the decline curve fraction (D) over the last segment, last rate, cumulative production, and estimated future WOR increase rate (fraction).

Time	Oil rate	N_p	N_p	WOR	Cut	Water Rate
y	BOPD	MMBOPY	MMBO	BW/BO		BWPD
0	4,300	60.3	60.3	11.5	0.92	49,450
1						

Table 7.2—Work table.

Analysis Procedure.

1. Set up table in 1-year increments. See data from Fig. 7.12 and Table 7.4.

2. Apply the last segment decline rate ($D = ?$) to forecast production for the Year 1 prediction, where $q_2 = q_1 \exp(-Dt)$ See example in **Fig. 7.14.**

3. Calculate cumulative oil production for Year 1.
$$N_p = \frac{q_1 - q_2}{D}.$$

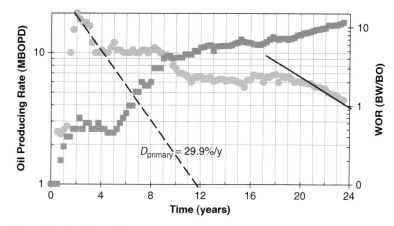

Fig. 7.14—Semi-log oil producing rate and WOR plots for the example field. Note the segmentation of the history.

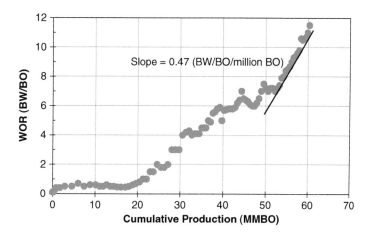

Fig. 7.15—WOR history applied to calculate slope of the line for last major segment. Adapted from Equity Sub-Committee (1964).

4. Cumulate oil production: See Table 7.4, the last recorded production value:

5. Rewrite the WOR slope equation to calculate the WOR at the end of Year 1.
$WOR_2 = m(N_{p2} - N_{p1}) + WOR_1$. See example in **Fig. 7.15.**

6. Revise the WOR equation to calculate new water production rate.
$q_w = q_o \cdot WOR$.

7. Calculate the equivalent water cut (f_w)
$f_w = \dfrac{q_w}{q_o + q_w}$.

In summary, the development allows one to reasonably extrapolate into the future expected changes in field performance and to predict the economics of expected production.

Fig. 7.14 shows the rate vs. time plot for the example field as well as the straight-line assessment of the last segment. WOR increase is the slope of the line.

As a matter of interest, before waterflood decline where $D = 39.9\%/yr$

Last segment oil rate where $D = 11.5\%/yr$

WOR increase – Slope of line, $m = \dfrac{y_2 - y_1}{x_2 - x_1} = \dfrac{11.4 - 6.1}{60.0 - 50.0} = 0.47 \dfrac{BW}{BO} / MMBO$.

Analysis Procedure.

1. Set up table in 1-year increments.
2. Apply the last segment decline rate ($D = 11.5\%/y$) to forecast production for the Year 1 prediction.
   $$q_2 = q_1 \exp(-Dt) = 4,300 \exp(-0.115 \cdot 1) = 3,833 \text{ BOPD}.$$
3. Calculate cumulative oil production for Year 1.x
   $$N_p = \frac{q_1 - q_2}{D} = \frac{(4,300 - 3,833) \text{ BOPD} * 365 \text{ days/yr}}{0.115 / yr} = 1.482 \text{ MMBO}.$$
4. Cumulate oil production: $60.0 + 1.48 = 61.48$ million BO.
5. Rewrite the WOR slope equation to calculate the WOR at the end of Year 1.
   $$WOR_2 = m(N_{p2} - N_{p1}) + WOR_1.$$
   $$WOR_2 = 0.47 \frac{\text{BW/BO}}{\text{MMBO}} (61.48 - 60.0) \text{MMBO} + 11.5 \frac{\text{BW}}{\text{BO}} = 12.2 \text{ BW/BO}.$$
6. Revise the WOR equation to calculate the new water production rate.
   $$q_w = q_o \cdot WOR = 3,833 \text{ BOPD} \cdot 12.2 \frac{\text{BW}}{\text{BO}} = 46,760 \text{ BWPD}.$$
7. Calculate equivalent water cut (f_w); $f_w = \dfrac{q_w}{q_o + q_w} = \dfrac{4,6760}{3,833 + 4,6760} = 0.92$.

q_o	t	N_p	N_p	WOR	q_w	f_w
BOPD	yr	BO/Y	MMBO	BW/BO	BWPD	
4300	0		60.0			
3833	1	1,482,616	61.5			
3416	2	1,321,554	62.8	11.5	49450	0.92
3045	3	1,177,988	64.0	12.2	46749	0.92
2715	4	1,050,019	65.0	12.8	43792	0.93
2420	5	935,951	66.0	13.4	40721	0.93
2157	6	834,275	66.8	13.9	37637	0.93
1922	7	743,645	67.6	14.3	34613	0.93
1714	8	662,860	68.3	14.7	31698	0.94

Table 7.3—Partial predicted depletion history.

Results of Analyses.

Fig. 7.16, below show oil flow rate and WOR predictions to be used in the economic assessments of the project.

Two Phase Flow and Buckley-Leverett. The following reviews the commonly applied fluid ratio equations. Produced fluid ratio plots reflect changing fluid flow rates for a two-phase system. Couple this observation with Darcy's law.

$$WOR = \frac{q_w}{q_o} = \frac{k_w B_o \mu_o}{k_o \mu_w B_w}$$

$$\text{or } OWR = \frac{q_o}{q_w} = \frac{k_o B_w \mu_w}{k_w \mu_o B_o}. \quad\quad\quad\quad (7.4)$$

The producing GOR, is the sum of the produced volume of solution gas and free gas.

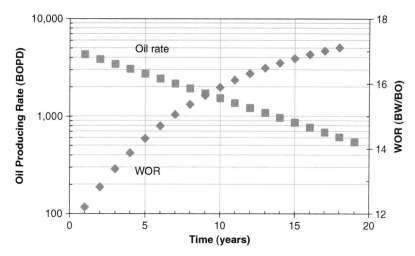

Fig. 7.16—Forecasting oil production and WOR rates. Note the dissimilarity of slopes of the lines.

$$GOR = \frac{q_g}{q_o} = R_s + \frac{B_o \mu_o k_g}{B_g \mu_g k_o}. \quad\quad (7.5)$$

The fractional flow of water term f_w, is often called water cut (WC) and defined is as

$WC = \frac{q_w}{q_t} = \frac{q_w}{q_o + q_w}$, where measurement is at surface conditions; q_w, q_o, and q_t are given in surface units (STB/D).

$$(STB/day) \quad\quad (7.6a)$$

$$WC = \frac{\mu_o B_o k_w}{\mu_w B_w k_o + \mu_o B_o k_w}, \text{ where the expression relates to the fraction of water flowing in the reservoir.} \quad (7.6b)$$

$$WC = = \frac{q_w}{q_t} = \frac{q_w}{q_o + q_w} = \frac{WOR}{WOR + 1}. \quad\quad (7.7)$$

Relate WOR to the fractional flow equation (f_w):

$fw = \frac{q_w}{q_t} = \frac{q_w}{q_o + q_w}$, q_w, q_o where and q_t are presented at subsurface conditions (RB/D).

$$WOR = \frac{f_w}{1 - f_w}\left(\frac{B_o}{B_w}\right), \quad\quad (7.8)$$

Relate water cut (f_w) to the relative permeability ratio.

$\frac{f_w}{1 - f_w} = \frac{k_w \mu_o}{k_o \mu_w} = C\frac{k_w}{k_o}$. When $C = \frac{\mu_o}{\mu_w}$,

$$\frac{k_w}{k_o} = \frac{f_w}{C(1 - f_w)}. \quad\quad (7.9)$$

Gas/oil or gas/water relationships can be derived in a similar manner.

Relative Permeability

Relative permeability, (k_o, k_g, or k_w) is a measure of the ability of a porous material to transmit a specific fluid, (i.e., oil, gas, or water) when saturated with other fluids. **Fig. 7.17** compares changes in the oil and water relative permeability curves when water is injected (imbibition) into an oil saturated core sample.

The curves are not straight lines, as a result of capillary and gravity affects. Note, water will not flow in the oil saturated core until water saturation increases to irreducible water, S_{wirr} = 30% value. Conversely, oil saturation becomes immobile at S_{or} = 26%.

A (k_w/k_o) ratio curve can be reasonably represented by two straight-line approximations.

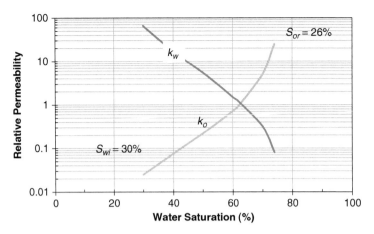

Fig. 7.17—Note that the straight-line approximations encompass the majority of the curve length. Eqs. 7.1 and 7.3 relate permeability to WOR or fraction of water flowing (f_w) field measurement data to the permeability values. Adapted from Prats (1961).

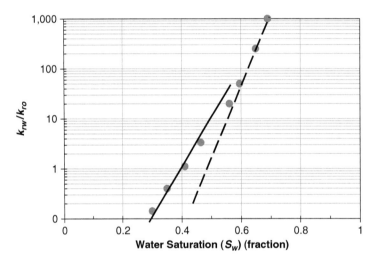

Fig. 7.18—Note that the straight-line approximations encompass the majority of the curve length. Eqs. 6.1 and 6.3 relate permeability to WOR or fraction of water flowing (f_w) field measurement data to the permeability values. Adapted from Prats (1961).

Frontal Advance. Frontal advance theory (Buckley and Leverett 1942) and field practice indicate that oil displacement by encroaching water occurs in two stages. **Fig. 7.19** shows the Buckley-Leverett saturation profile for an oil reservoir with the injection well located on the left and producing well on the right. The displacement process occurs in two stages.

1. Initial (primary) stage—constant rate with a zero WOR that lasts until breakthrough of the constant saturation front to the producing well at X3. The constant saturation front represents the stabilized zone. Laboratory studies have shown that the stabilized zone saturation remains constant. Therefore, the (df_w/dS_w) slope must remain constant.
2. Secondary and subordinate stage—characterized by a constantly increasing water saturation. The shape of the transitional curve approximates the relative permeability curves.

The foundation to the Buckley-Leverett solution is the fractional flow of water vs. change in water saturation curve (f_w vs. S_w) plot, an example of which is shown in **Fig. 7.20**. The curve shows how two phase flow permeability (k_o, k_w) affects the fraction of water flowing in a system. The curve begins at S_{wi} and ends at S_{or} and reflects the flood-out history of a waterflooded core.

Transfer the previously discussed Buckley-Leverett concepts to a fractional flow curve, shown in **Figs. 7.21a, b** in order to predict future performance from relative permeability data.

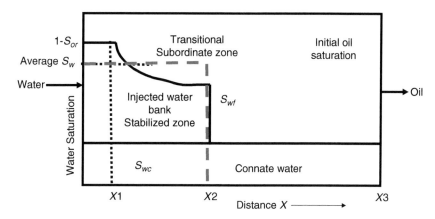

Fig. 7.19—A schematic of an injected water front moving toward a producing well at X3. (S_{wf}) represents the water saturation at the front. ($\bar{S}_{wf}$) Represents the average saturation between 0 and X2. Movable oil saturation = $S_{wi} - (1-S_{or})$. Adapted from Buckley and Leverett (1942).

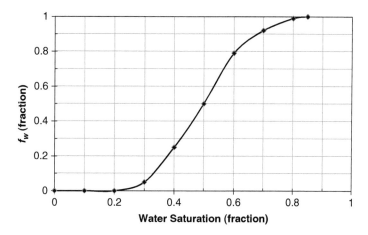

Fig. 7.20—The fractional flow curve reflecting the waterflood desaturation history of oil and water, the two phases in a waterflooded core.

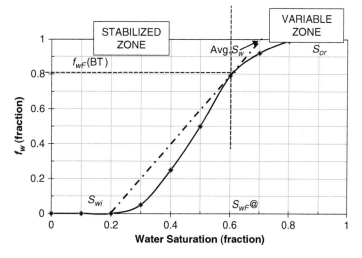

Fig. 7.21a—The straight-line extrapolation divides the (s-shaped) into stabilized and variable saturation zones.

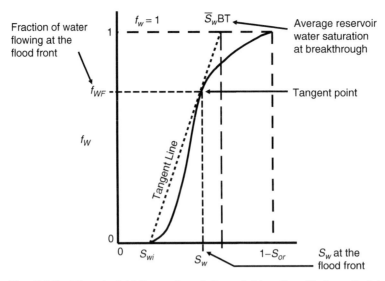

Fig. 7.21b—The straight-line extrapolation divides the (S-shaped) into stabilized and variable saturation zones. Points of interest on the fractional flow curve are f_{wFBT} – fraction of water flowing at breakthrough, S_{wF} – water saturation at flood-front breakthrough, $\overline{S}_{wBT}$ – average water saturation at breakthrough, S_{or} – residual oil saturation, S_{wi} – initial water saturation.

Fundamentals of analytic geometry dictate the slope of a secant line intersecting a curve at two or more points represents the average slope of the curve. Therefore, the (df_w/dS_w) slope at the time of water breakthrough is represented by the straight-line drawn tangent to the fractional flow curve.

The fractional flow curve has now been divided into stabilized and variable saturation zones. Recall the secant angle is equal to 1/cosine. **Fig. 7.21b** shows the relation of the secant line to the fractional flow curve.

A relative permeability vs. cumulative production relationship can be established at any point along the segment as long as one stays within the two endpoints.

Observations. The observations are

- The point of tangency defines the fraction of water flowing (f_{wBT}) at the time of breakthrough and the water saturation (S_{wF}) at the front.
- Extending the straight line to $f_w = 1$ defines the average water saturation behind the flood front at breakthrough.
- The displacement concept derives a stabilized and a variable saturation zone.
- A straight line initiated at the (f_w) axis and drawn to a point of tangency on the fractional flow curve will result in a water saturation (S_{wFT}) measured at the front.
- Extending the line to the $f_w = 1$ value results in an average water saturation behind the flood front.
- Primary-stage-shock front—represents oil volume displaced when water breaks through at the producing well minus (S_{wF}).
- Subordinate-stage involves transition zone—water displacing oil controlled by relative permeability effects [S_{wF} to $(1-S_{or})$].

Creating Pseudorelative Permeability Curves. Field WOR or fraction of water flowing (f_w) values reflect the sum total of

- Rock-properties variation
- Field architecture effects
- Well locations
- Completions
- Production and injection rates
- Difference in density of displacing fluid
- Density of displaced fluid that causes overriding (when density of displacing fluid is lower than density of displaced fluid) or underriding (when density of displacing fluid higher than density of displaced fluid). Gravity affects the vertical efficiency not only in heterogeneous reservoirs but also in homogeneous reservoirs. Gravity effects will be important when vertical communication is good and gravity forces are larger than viscous forces.

Field studies often include relative permeability effects from laboratory derived core plug experiments or rather arbitrary adjustments from history matching two- or three-phase-flow field production. The following discussion relates field or well production characteristics to develop comparable relative permeability ratio curves.

Representing (k_w/k_o) Expression with a Straight Line. Welge (1952) defined the fraction of water flowing (f_w) in terms of the ratio of relative permeability and viscosity.

$$f_w = \frac{1}{1 + \frac{k_o \mu_w}{k_w \mu_o}} \quad \quad (7.10)$$

Assume a straight-line approximation over the late time portion of the (k_o/k_w) curve and rearrange to the equation of a straight line.

$$\ln \frac{k_w}{k_o} = mS_w + \ln a. \quad \quad (7.11a)$$

Substitute cumulative oil production for the water saturation.

$$\ln \frac{k_w}{k_o} = mN_p + \ln a. \quad \quad (7.11b)$$

Change to the exponential form.

$$\frac{k_w}{k_o} = a \cdot \text{Exp}(m \cdot Np). \quad \quad (7.11c)$$

The two outcome coefficients representing the straight-line approximation of the (k_o/k_w) curve are slope (m) and intercept ($\ln a$) terms. Recall Eq. 7.6 relates watercut (f_w) to the relative permeability ratio (k_o/k_w).

$$\frac{k_w}{k_o} = \frac{f_w}{C(1-f_w)}, \text{ where } C = \frac{\mu_o}{\mu_w}. \quad \quad (7.12)$$

In conclusion, we can say that portions of the ($\ln \frac{k_w}{k_o}$ vs. N_p) curve can be represented by the equation of a straight line.

Therefore, relative permeability can be related to production segments defined by a series of straight lines.

Example Field Analyses. The following example illustrates application of the previously discussed principles to solve actual field problems. Refer to **Table 7.4.**

B_o = 1.05 RB/STB,
B_w = 1.0 BW/STB,
m_w = 1.0 cp
m_o = 9.3 cp
φ = 33%,
S_{wi} = 23%,
S_{or} = 31%,
S_{gr} = 8%,
OOIP = 1,877 STB/acre-ft, = 157 MMBO,
Movable oil = 1,360 STB/acre-ft, = 114 MMBO
N_p = 60,318,000 BO credited to date,

Table 7.4—Field parameters. Adapted from Equity Sub-Committee (1964)

Review Performance. The field has been waterflooded for a number of years. **Figs. 7.22** and **7.23** show the 23-year history of oil, gas, and water production, and the number of producing wells over the life of the field.

History is not a smooth curve but can be divided into a series of straight lines, each denoting periods of constancy. These same segments can be correlated with the oil rate plots.

Apply **Fig. 7.24** to translate time units into cumulative production units when needed. Tangent straight lines representing operating segments of the Buckley–Leverett plot are drawn in **Fig. 7.25** in a (f_w vs. N_p) plot.

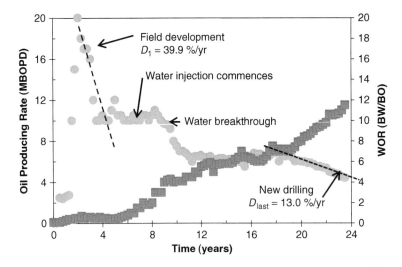

Fig. 7.22—Oil production rate and WOR for example field. Majority of field drilling essentially ceased by Year 6. Note the segmented structure of the field history. Adapted from Equity Sub-Committee (1964).

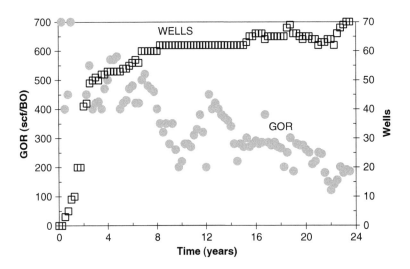

Fig. 7.23—GOR and drilling history. Declining GOR was probably the result of continuous loss of reservoir pressure because of field development. Majority of field drilling essentially ceased by year 6. Are there any similarities between the two previous figures, Figs. 21 and 22? Adapted from Equity Sub-Committee (1964).

Interpretation of Field History. The interpretation of field history is given below.

- How many production segments can you identify?
  Segment 1 and 2 – Primary development and initiation of waterflood.
  Segment 3 – Major water breakthrough at the end. (N_p at 30 million BO).
  Segments 4 and 5 – Increasing WOR. Water cycling recently increased significantly.
- Well number remained relatively constant at 62 from Year 7 to Year 22 when an infield drilling campaign commenced.
- Depletion and watering out of the major pay sands caused oil production to decline to approximately 4,100 BOPD. Water injection rates have varied from 20,000 to 40,000 BWPD.
- Pressure depletion conditions were evident with production falloff and a dramatic GOR decline. The $D_{initial}$ = 39.9%/yr decline provided additional evidence for why the waterflood was installed.
- How does the continual GOR reduction affect oil viscosity and inflow performance?

Segment parameters are collected in **Table 7.5**. Figure 7.24 shows the last segment oil rate where $D = 13.0\%$/yr.

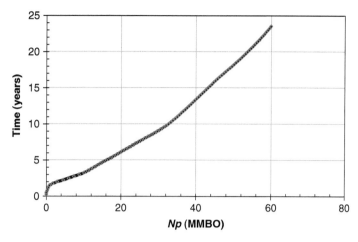

Fig. 7.24—Comparing historical time to cumulative production. The plot is useful to relate cumulative production to time values.

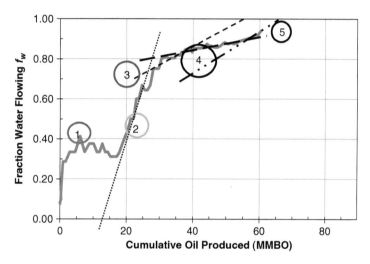

Fig. 7.25—Rate vs. time and WOR history divided into five producing segments. The slope of segment 1 represents the initial field development period is somewhat problematic, and is not really germane to the discussion. Table 7.6 lists the field parameters for the segments. Adapted from Equity Sub-Committee (1984).

	f_w range	WOR	Interval	Cumulative Production	k_w/k_o
	Frac. Water		years	MMBO	
1			0 – 5.0	0.0 – 17.0	
2	0.32 – 0.65	0.47 – 1.9	5.0 – 8.0	17.5 – 30.0	0.05 – 0.20
3	0.78 – 0.84	3.0 – 5.3	8.5 – 9.5	33.0 – 40.0	0.25 – 0.57
4	0.84 – 0.89	5.3 – 8.1	11.0 – 15.5	40.0 – 51.0	0.57 – 0.87
5	0.90 – 0.92	9.0 – 11.5	16.3 – 19.0	57.0 – 59.0	0.97 – 1.24

Table 7.5—Segment components.

Primary and Subordinate Stages. The primary stage comprises Segments 1 and 2. The initial stage reflects drilling up the field in a geological sense. Pressure-depletion conditions were observed, and water injection commenced at Year 5.

Pronounced water breakthrough occurred at 32 MMBO Year 9) at a pronounced fractional flow that increased during Stage 2. The water breakthrough event spanned Segment 3 as water production gradually increased. Inclusion of water in the producing stream caused significant falloff of the oil production rate.

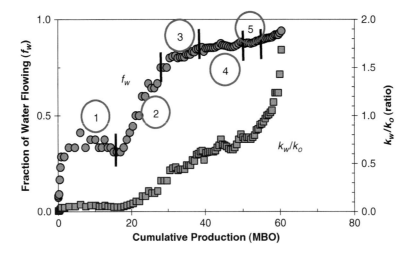

Fig. 7.26—Comparing fractional flow and relative permeability curves. Note how the relative permeability curve accentuates water flowing within the producing system. The historical segments are shown on the (f_w) curve. Adapted from Equity Sub-Committee (1984).

The subordinate stage shows $f_w > 80\%$. Oil produced at breakthrough when $N_p = 32.0$ MMBO and Recovery = 18.1 %. This value is equivalent to the change in water saturation ($\bar{S}_w$) value measured behind the flood front on the Buckley-Leverett curve. This is represented with Segments 4 and 5, where Segment 5 is associated with (probably unsuccessful) infill drilling.

Fractional Flow Curves. Apply Eq. 7.5 to compare fractional flow of water (f_w) to an equivalent (k_w/k_o) curve. Plot the results as a function of cumulative oil produced shown in **Fig. 7.26.** The slope of the last segment for the (k_w/k_o) curve is more pronounced, which indicates an abrupt increase of water channeling through the depleted sand members.

Extrapolation of the straight line to the fractional flow ($f_w = 1$) value results in an estimated ultimate recovery (EUR) value.

The following observations can be made:

- The historical divisions noted in segment plot are accentuated in the (k_w/k_o) plot.
- The last few data points seem to indicate that the very recent drilling campaign was detrimental to production operations.

Alternative Interpretive Concepts. Ershaghi and Abdassah (1984) combined the fractional flow equation and Buckley-Leverett displacement concepts to develop a method for

- Predicting performance,
- Determining a segment or pseudorelative permeability ratio curve for a developed field

Two major assumptions form the basis for the method:

- A linear (1D), homogeneous system and constant operating conditions.
- The reservoir can be fitted to a geometric outline and the (k_o/k_w vs. N_p) plot can be represented by straight line approximations.

The authors state that the method implies that reservoir and well configurations and operations do not change over the life of the prediction.

Welge (1954) developed an equation describing water saturation at any point in the system as a function of the fraction of water flowing (f_w). The equation is presented as the equation of a straight line.

$$E_R = \frac{1}{b(1-S_{wi})}\left[\ln\left(\frac{1}{f_w}-1\right)-\frac{1}{f_w}\right]-\frac{1}{(1-S_{wi})}\left[S_{wi}-\frac{1}{b}\ln(A)\right]-\left[\ln\left(\frac{1}{f_w}-1\right)-\frac{1}{f_w}\right], \quad \ldots \ldots \ldots \ldots (7.13)$$

where $E_R = \dfrac{N_p}{N}$, the fractional recovery, "y" plotting value; f_w = portion of "x" plotting value; $A = a\dfrac{\mu_w}{\mu_o}$, intercept of the straight line; and (a and b) define the location of the straight line.

Fig. 7.27 shows an example plot.

Rearranging to the equation of a straight line and including relations for f_w, WOR, and $S_w - N_p$ result in a rearranged version of the Welge development:

"y" plotting variable: $(\ln OWR - OWR - 1)$	Outcome: Slope $n = m(1 - S_{wi})$
"x" plotting variable: N_p	Intercept: $c = mS_{wi} - \ln a\left(\dfrac{\mu w}{\mu_o}\right)$

Table 7.6—Plotting variables

The Welge plot highlights the operating segments of the example field with Segments 1 and 2 spanning field depletion, Segments 3 and 4 representing waterflood displacement, and finally Segment 5 showing field exhaustion (**Fig. 7.28**).

Plotting $\left(\dfrac{1}{\text{WOR}} \mp \text{WOR vs. } N_p\right)$ Lijek (1989) showed that plotting $\left(\dfrac{1}{\text{WOR}} + \text{WOR vs. } N_p\right)$ produced essentially the same results as those by Ershagi. **Fig. 7.29** shows the plot for the example field.

Multiple Performance Plots

This exercise applies multiple rate plots from Well 209 to help analyze and interpret performance. Reinforcing conclusions developed from other interpretations can enhance the quality of the final outcome. The discussion is best described in an example, adapted from Bondar and Blasingame (2002).

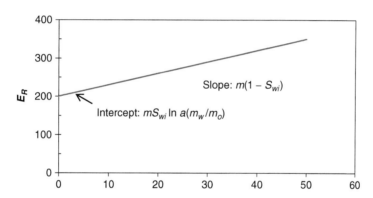

Fig. 7.27—Interpretation of the Ershagi and Omorigi (1954) equation of a straight line.

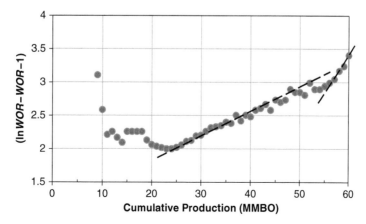

Fig. 7.28—The Ershagi idea works well for the example field, particularly when dividing a performance history into segments. Adapted from Equity Sub-Committee (1984).

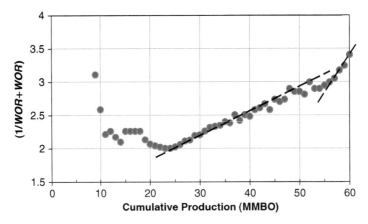

Fig. 7.29—The Lijek plot $\left(\dfrac{1}{WOR} \mp WOR \text{ vs. } N_p\right)$ reflects essentially the same nature as the Ershagi plot. Adapted from Equity Sub-Committee (1984).

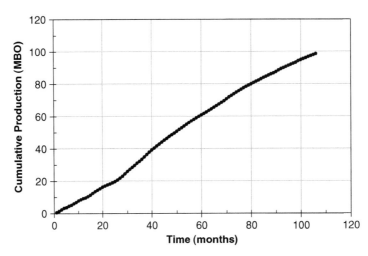

Fig. 7.30—Relating cumulative production to producing time. Adapted from Bondar and Blasingame (2002).

Well history is summarized in the following:

- Well 209 was completed, acidized, and fractured in 1987.
- Waterflooding commenced at Month 24.
- A workover in Month 60 repaired a hole in the tubing.
- The well produces from two sands of differing permeability.

Various rate vs. time or rate vs. cumulative recovery plots listed in the following figures are provided for interpretation to aid when forming an opinion concerning performance history, reserves, and future performance:

- **Fig. 7.30**—Relating time to cumulative production plot—used during interpretation.
- **Fig. 7.31**—Composite of all plots; review possible plots.
- **Fig. 7.32**—Cartesian rate vs. time production history—helps interpreting well history.
- **Fig. 7.33**—Semi-log rate vs. time production history; find any straight lines in the history.
- **Fig. 7.34**—Rate-cumulative recovery; if not absolutely straight be cautious.
- **Fig. 7.35**—Oil and water production histories. How do the curves relate to each other?
- **Fig. 7.36**—WOR history—compare to oil and water rates.
- **Fig. 7.37**—Total production rate history—compare to oil and water rates.
- **Fig. 7.38**—Ershagi plot—compare Fig. 7.38 to Fig. 7.40. Which figure do you find to be most expressive?
- **Fig. 7.39**—Watercut plot.
- **Fig. 7.40**— $\dfrac{1}{WOR} + WOR$ vs. N_p.

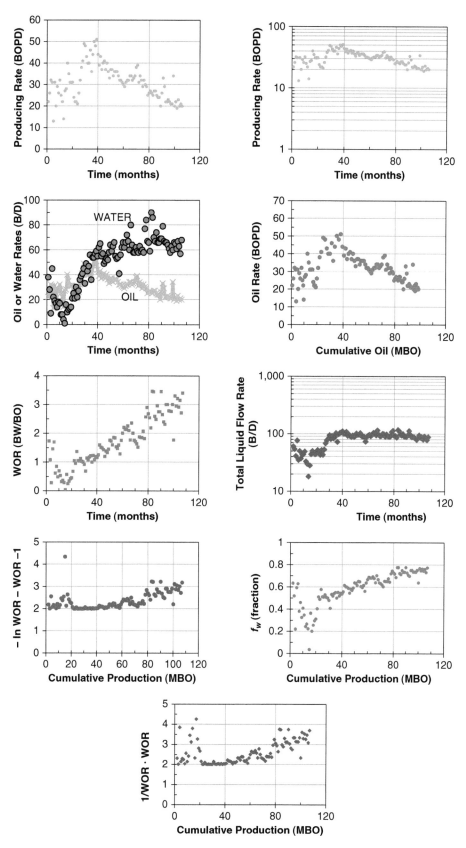

Fig. 7.31—Composite performance history. Adapted from Bondar and Blasingame (2002).

138 Analysis of Oil and Gas Production Performance by Poston, Laprea-Bigott, Poe

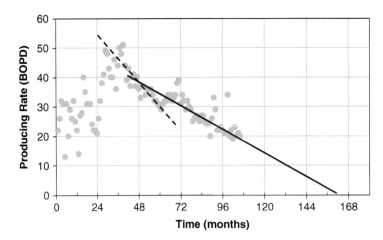

Fig. 7.32—Cartesian rate vs. time plot for Well 209. Adapted from Bondar and Blasingame (2002).

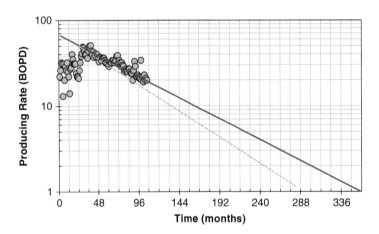

Fig. 7.33—Semi-log rate vs. time plot for Well 209. Adapted from Bondar and Blasingame (2002).

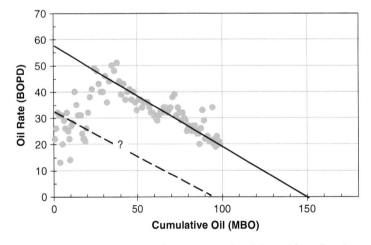

Fig. 7.34—Oil rate vs. cumulative recovery plot. Adapted from Bondar and Blasingame (2002).

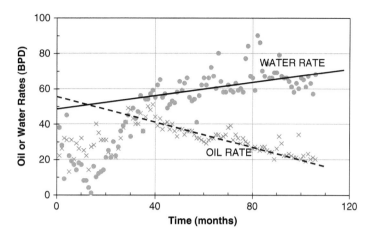

Fig. 7.35—Production was very erratic until about month 30; after that, there was good steady flow and a constant decline rate. Adapted from Bondar and Blasingame (2002).

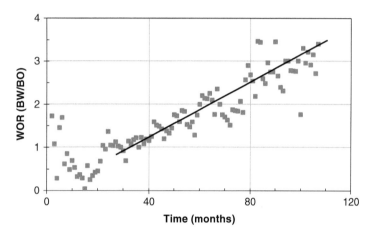

Fig. 7.36—A Cartesian scale WOR vs. time plot. Note the good straight-line fit of the data beginning at the same time as the onset of the exponential decline observed in Fig. 7.33. Data scatter seems to increase after 70 months. Why? Adapted from Bondar and Blasingame (2002).

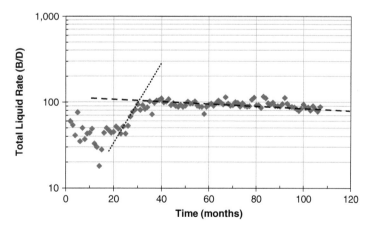

Fig. 7.37—Total well liquid flow rate. Is the well capacity of 100 BFPD caused by the mechanical completion arrangement? Adapted from Bondar and Blasingame (2002).

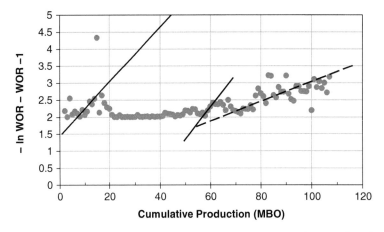

Fig. 7.38—The Ershagi plot. Adapted from Bondar and Blasingame (2002).

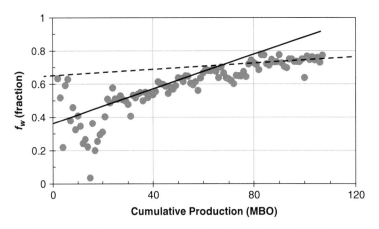

Fig. 7.39—Watercut plot. Adapted from Bondar and Blasingame (2002).

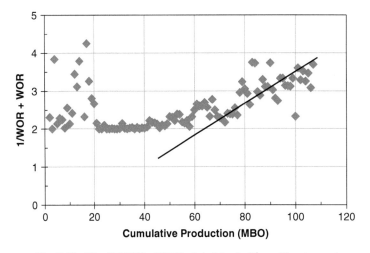

Fig. 7.40—The (1/WOR + WOR) plot. Adapted from Bondar and Blasingame (2002).

The following is a composite of all the production curves included in the study.
Can you identify the period:

- Over which pressure-depletion production occurred?
- When water injection commenced?
- When water breakthrough occurred?
- When the hole in the tubing was fixed?

Do the straight lines have a mathematical significance? Can you predict when the well will produce 5 BOPD? Does this straight line fulfill the basic concept of the Arps exponential decline definition?

Estimate the following:

- The EUR
- Remaining reserves, if the economic limit = 5 BOPD
- Primary reserves
- Secondary reserves

Next, calculate the decline rate. Was the workover successful?
Do you believe a straight line provides a reasonable primary reserves estimate? Predict the following:

- The EUR
- Primary and secondary reserves when the economic limit = 5 BOPD

Predict the expected water production rate at month 120.
Water production gradually increased from a low of 0.3 BW/BO in Month 17 to 3.2 BW/BO. Should one be concerned about water handling costs at this point?
What is the WOR at Month 120?
Can you come to any conclusions about the total flow rate? What would happen if larger tubing were installed? Total production appears to be a function of the lifting ability of the well. Is scaling a problem?

Fig. 7.35 is a very important plot because it indicates the mechanical arrangement should be reviewed to determine if the total flow rate could be increased.

How many segments can the plot be divided into? Can you identify when water breakthrough event occurred? Is this plot very descriptive?

Apply Eq. 7.4 to calculate WOR when $f_w = 0.6$.

Compare the WOR, Ershagi, and Lijack plots for goodness of interpretation. Which seems to be the best?

What is the relationship between fractional flow, WOR, and production? Do you note any differences in the relations?

f_w	WOR
(fraction)	(RB)
0.4	0.87
0.65	2.41

Table 7.7—Fractional flow and WOR results.

$N_{p(primary)} = 19{,}050$ BO, $N_{p(secondary)} = 79{,}680$ BO, $B_w = 1$; $B_o = 1.3$:

$$WOR = \left(\frac{f_w}{1-f_w}\right)\frac{B_o}{B_w} = \left(\frac{0.4}{1-0.4}\right)\frac{1.3}{1} = 0.87.$$

$$WOR = \left(\frac{f_w}{1-f_w}\right)\frac{B_o}{B_w} = \left(\frac{0.65}{1-0.65}\right)\frac{1.3}{1} = 2.41.$$

Summary of Analysis.

In summary, the rate vs. time plots held steady-state conditions over 30 months. The oil production rate declined exponentially at 15.5%/yr. Water production slowly increased. For the WOR vs. time plot, WOR increased exponentially at a fairly uniform 19.7%/yr.

- For oil and water vs. time plot, there was a constant total liquid flow indicating that we perhaps should check to see if we could install a larger pump.

Conclusions. In conclusion, interpretation of multiple performance plots can furnish valuable clues to waterflood performance. Performance plots can be constructed to determine oil in place and maximum movable oil (**Fig. 7.34**), but care should be exercised in the quantitative meanings of this interpretation.

Well Diagnostics Plots

Chan (1995) stated that excessive producing fluid ratios at the wellbore can be divided into coning and channeling cases.

Water cones upward to the perforations from an underlying water leg or gas cusps downward to the oil zone perforations. Introduction of the second phase into the well stream can occur any time after production has been initiated. Generally, water or gas producing rates increase smoothly as a function of time because of the growth of the cone.

Water or gas channels from a high-permeability layer included in the producing zone. Breakthrough of water or gas is a function of the absolute permeability of the rock layers and the mobility ratio.

Near-wellbore channeling often occurs because of behind pipe communication caused by a bad cement job. WOR or GOR abruptly increases for this case.

Start interpretation of the unwanted fluid can only be achieved when the following information is available:

- A log profile and knowledge of reservoir characteristics and drive mechanism(s)
- Gas, oil, and water production and flowing pressure history
- A schematic of the completion setup

Simulated relationship between coning and channeling on a logarithmic WOR vs. time plot is shown in **Fig. 7.41.** Adapted from Chan (1995).

Note the smoothly increasing WOR for the coning situation, by means of the viscous forces gradually overpowering gravity forces. On the other hand, water channeling occurs in a much more dramatic manner once breakthrough to the wellbore occurs. A performance history can be divided into three different periods:

- Flat-lying production until water encounters the perforations at some departure time, which can occur anytime during the production history.
- Increasing water production. Water production for the coning case smoothly increases as a function of time, while behind-pipe channeling can occur quite dramatically. Water channeling from a multilayered producing zone does not occur in so dramatic a manner.
- The third part of the history occurs when water production overrides oil production and the well becomes uneconomic because mostly water is being produced.

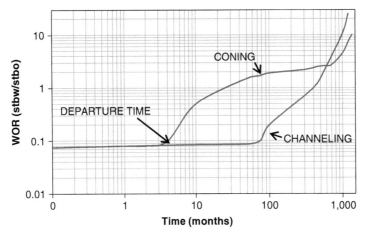

Fig. 7.41—Typical (WOR) performance plots define differences between coning and channeling. Adapted from Chan (1995).

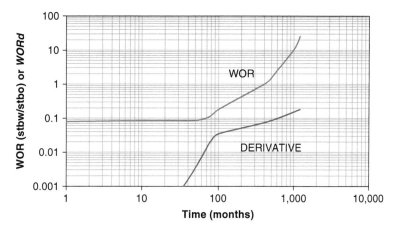

Fig. 7.42—WOR and derivative curves for the multilayer water channeling situation. Adapted from Chan (1995).

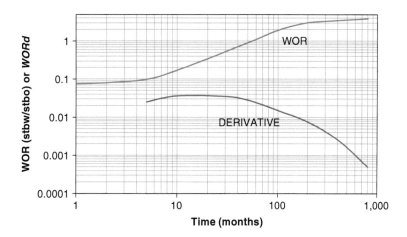

Fig. 7.43—WOR and derivative curves for the water coning situation. Adapted from Chan (1995).

Derivative Plots. Derivative plots can help determine the origin of the unwanted fluid entering the production stream. Chan (1995) included a derivative curve to aid in the interpretation. **Fig. 4.3** shows the WOR and time derivative relationship for the water channeling case. Note the dramatic increasing slope for the water channeling derivative curve.

Note that the beginning of water breakthrough at the point of departure causes a dramatic upward shift of the WOR curve with a consequent increase of the derivative or slope indicator curve.

On the other hand, a comparative plot for the water coning situation (**Fig. 7.43**) shows a smoothly increasing WOR while the derivative plot reflects a diametrically increasing slope.

Fig. 7.44 shows the effects of water coning and then channeling on a well. Note the change in direction of the derivative curve for this case.

It is important to note that in real life these interpretations should be attempted in conjunction with the performance histories, well schematics, and completion report.

Additional Requirements. The additional requirements are

- Knowledge of reservoir geology; refer to Figs. 7.1 and 7.2.
- A plot of the production history.
- Linear or semi-log WOR or gas/water-ratio vs. cumulative production and water cut vs. cumulative production plots to evaluate recovery efficiency.

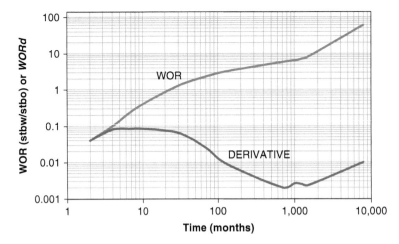

Fig. 7.44—WOR and derivative curves for the gas coning and then channeling situation. Adapted from Chan (1995).

Example Problem 1. The following Well 20 example problem occurred in an offshore well completed in a high permeability sand. Water production was negligible.

Oil and gas producing rates remained stable at approximately 500 BOPD and constant GOR for approximately 42 months. The rate was restricted below the potential because of the proximity to the gas cap. The gas producing rate more than tripled after Month 46 (**Fig. 7.45**).

Fig. 7.46 shows GOR and GOR derivative plots. The derivative plot was generated with the slope equation, which is easily set up on a spread sheet. An example calculation is shown here:

$$m = \frac{y_2 - y_1}{x_2 - x_1} = \frac{GOR_2 - GOR_1}{t_2 - t_1}; \text{ for } t = 8m, = \frac{667 - 630}{9 - 8} = \frac{37 \text{scf}}{BO - m}$$

The GOR remains essentially constant up until approximately Month 50 when there is evident gas breakthrough. Increased gas saturation at that time reduced the oil producing ability. Did the threefold rate increase cause the gas-cap gas to enter the perforations?

There is considerable spread in the calculated derivative values. In fact, 30% of the values were negative because of the erratic nature of the data. However, one perceives that the derivative trend increases at about the same time that the producing rate increased. Gas-cap breakthrough was probably the cause for the sudden increase in GOR, even though the derivative curve suddenly increased. This technique probably works best for water because of the high viscosity of water.

Summary of Analysis. The technique should be a useful tool to help analyze well problems. Each field should have characteristic curves depending on the reservoir-rock and fluid characteristics. Simulation studies generating the curves shown at the start of the discussion are considerably smoother than field data, but the principles are evident.

Example Problem 2. The producing history for Well MR-321 which has been water flooded for a number of years, is shown in **Fig. 7.47**. The oil producing rate has continued to decrease as a function of time after about 40 months of production, while the water rate has remained essentially constant. Please answer the following questions.

1. Water injection commenced at Month 0. Is there evidence of waterflood effect?
2. Can you determine the time of water breakthrough to Well MR-321?
3. Is the decline exponential in nature?
4. There has been a history of holes in the tubing. Can you spot these instances?
5. Does the derivative plot furnish evidence that a hole in the tubing caused the recent increase in water production?

Fig. 7.48 shows the derivative plot.

Example Problem 3. Evans Well 11 produces from a mature waterflood of a layered sand. Average permeability values for the individual members range from 100 to 40 to 1 md, respectively. Water production has become excessive, and project viability is of concern.

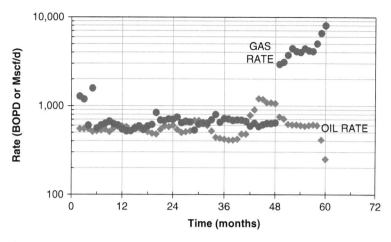

Fig. 7.45—Producing history for Well 20. Note the sudden onset of increased gas production at 46 months. Could the increased oil rate have brought the gas/oil contact down to the perforations?

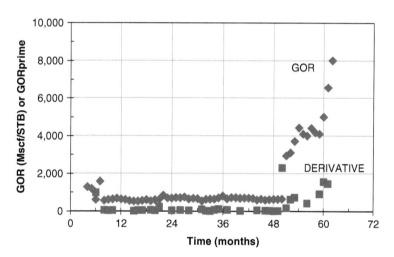

Fig. 7.46—GOR and derivative history for Well 20.

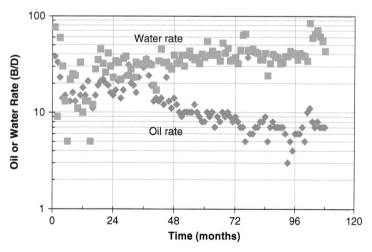

Fig. 7.47—Producing history for Well MR-321.

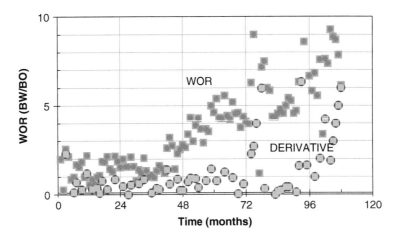

Fig. 7.48—Producing WOR history for Well MR-321, including WOR and derivative curves.

Study the following production figures to determine the number of layers that are probably flooded out and estimate remaining performance for Evans Well 11.

Learning Objective. The learning objective is to see how knowledge of geology and field operations can aid when estimating future performance.

Fig. 7.49 plots the well WOR history. Note the difference between the WOR plateaus for water breakthrough of the two sands.

Fig. 7.50 presents the (k_w/k_o) ratios calculated from the production data.

Answer—Example Problem 3. It seems the highest-permeability layers have flooded out and there is little remaining potential remaining in the waterflood.

Example Problem 4. Well MR-321 was drilled and completed in a shallow oil field undergoing a waterflood. A pump jack was installed. The well is continually pumped off. The producing interval is composed of three differing-permeability sand members.

Learning Objective. The learning objective is to couple known production history and information with reconstituted production plots to forecast future performance. Study the following performance plots to answer the following questions.

1. How many production segments are there?
2. Compare the Cartesian to the semi-log plot. Which do you think furnishes the most information?
3. When did water breakthrough occur for the first layer and second layer?
4. Is an exponential decline apparent?
5. Is there evidence of another water breakthrough event?
6. What would happen if a larger pump were installed?
7. Did the scale cleaning in Month 96 help production?

The plots are shown below in **Figs 7.51 through 7.55. Fig. 7.51** shows the cartesian rate vs. time. **Fig. 7.52** shows the log rate vs. time. **Fig. 7.53** shows the rate vs. cumulative production. **Fig. 7.54** shows the oil and water flow rates. **Fig. 7.55** shows the total flow through the system.

Answers—Example Problem 4.

1. How many production segments are there? Three: pressure depleting, water injection kick, and water cycling.
2. Compare the Cartesian plot to the semi-log plot. Which do you think furnishes the most information? Cartesian.
3. When did water breakthrough occur for the first layer and second layer? Probably near Month 40 and to date.
4. Is an exponential decline apparent? Yes, near Month 48, $D = 19.8\%$.
5. Is there evidence of another water breakthrough event? Possibly at 100 months.
6. What would happen if a larger pump were installed? Possibly an increased rate, as shown in **Figs. 7.54 and 7.55.**
7. Did the scale cleaning in Month 96 help production? Yes.

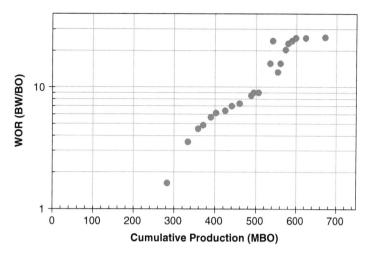

Fig. 7.49—The logarithmic WOR for the Evans Well.

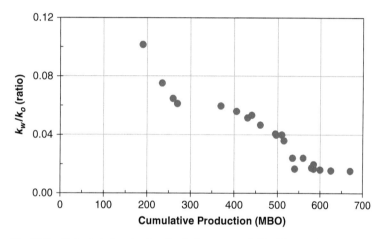

Fig. 7.50—Calculated water to oil relative permeability history for the Evans Well. Refer to Eq. 7.1.

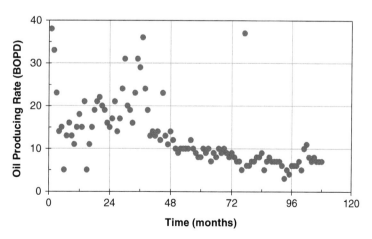

Fig. 7.51—Cartesian rate vs. time plot. Production leveled out after breakthrough. Why?

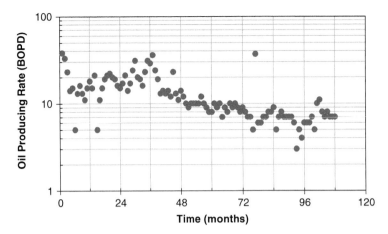

Fig. 7.52—Log rate vs. time.

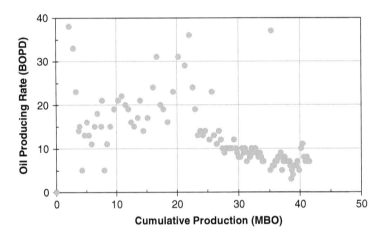

Fig. 7.53—Rate vs. cumulative production. Note the degree of scatter over the first half of the history. Can you realistically extrapolate to a meaningful EUR estimate from the plot?

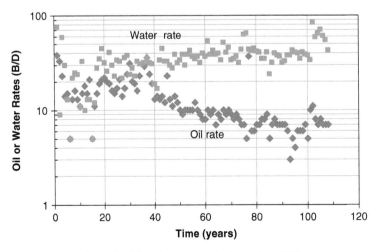

Fig. 7.54—Oil and water producing rates, BPD.

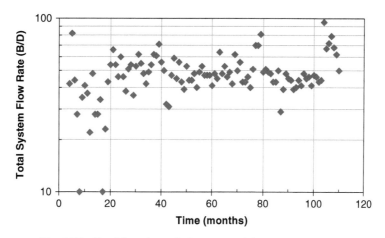

Fig. 7.55—Total flow through the system (oil and water), in B/D.

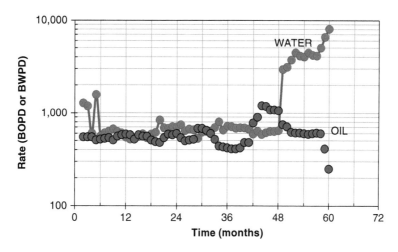

Fig. P7.1.1—Rate vs. time history for PP4 well. Baker and Anderson (2003).

PROBLEMS

Example Problem 7.1. Produced water in Well PP4. **Fig. P7.1.1** shows the oil and water producing rate history for the PP4 well. Note that water was the predominant produced fluid in two instances. What could be the cause for these two cases? Lost completion fluids, water coning, hole in tubing, or packer leak?

Learning Objective. Analyze production history.

Example Problem 7.2. The MR321 well was drilled and completed in a shallow oil field undergoing a waterflood. A pump jack was installed. The well is continually pumped off. The producing interval is composed of three sand members of differing permeability.

Learning Objective. Couple known production history and information developed from production plots to forecast future performance.

Study the performance plots in **Figs. P7.2.1 through P7.2.5** to answer the following questions.

1. How many production segments are there?
2. Compare the Cartesian to the semilog plot. Which plot do you feel furnishes the most information?
3. When did water breakthrough occur for the first layer and second layer?
4. Is an exponential decline apparent?
5. Is there evidence of another water breakthrough event?
6. What would happen if a larger pump were installed?
7. Did the scale cleaning in Month 96 help production?

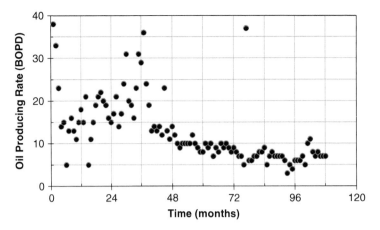

Fig. P7.2.1—Cartesian rate vs. time plot. Production really leveled out after breakthrough. Why? Baker and Anderson (2003).

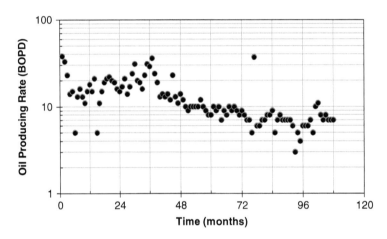

Fig. P7.2.2—Plot of ln rate vs. time. Baker and Anderson (2003).

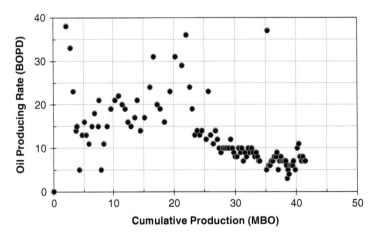

Fig. P7.2.3—Rate vs. cumulative production. Note the degree of scatter over the first half of the history. Can you realistically extrapolate to a realistic EUR estimate from the plot? Baker and Anderson (2003).

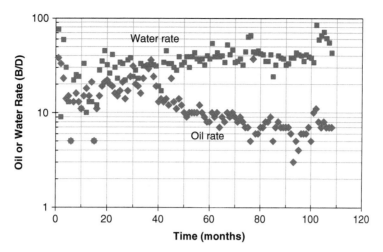

Fig. P7.2.4—Oil and water producing rates (B/D). Baker and Anderson (2003).

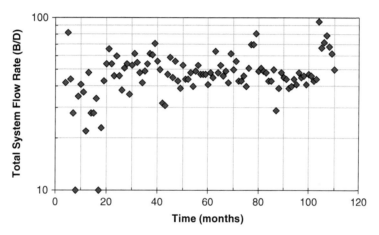

Fig. P7.2.5—Total flow through the system (oil + water) in B/D. Baker and Anderson (2003).

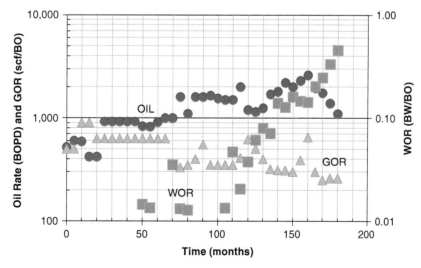

Fig. P7.3.1—Performance history of ME 232 well.

Example Problem 7.3. The ME 232 well is a part of a large ongoing pressure-maintenance project. **Fig. P7.3.1** relates the 15-year oil, water and gas producing history. Please note the trends of the production histories.

Source water has been a constant problem. There is some evidence of possibly not attaining sufficient water injection requirements to effectively deplete the reservoir. **Fig. P7.3.1** shows the well performance history.

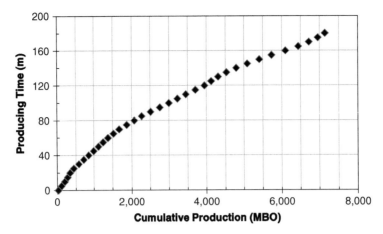

Fig. P7.3.2—Relating cumulative production to producing time. Baker and Anderson (2003).

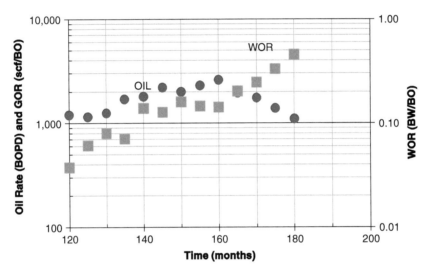

Fig. P7.3.3—Estimation of oil decline rate. Baker and Anderson (2003).

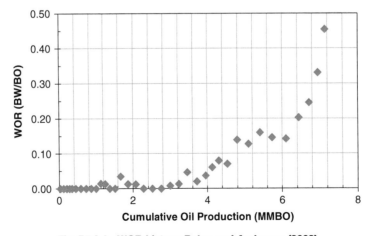

Fig. P7.3.4—WOR history. Baker and Anderson (2003).

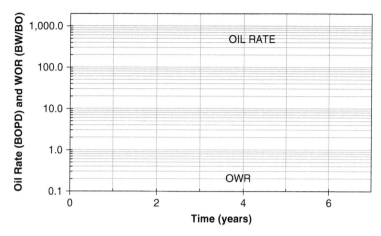

Fig. P7.3.5—Performance prediction for ME 232 well. Baker and Anderson (2003).

Learning Objective. To analyze production history by segments and predict future performance.

Analysis Procedure.

- Divide the producing history into segments, each representing individual operating units as the first step of your analysis. Fill in the blanks in **Table P7.3.1.**
- Construct the producing time vs. cumulative production plot (**Fig. P7.3.5**) to determine time vs. production relationships.
- Construct a semilog rate vs. time plot (**Fig. P7.3.2**) to estimate future oil rate performance.
- Construct a WOR vs. cumulative-production plot (**Fig. P7.3.3**) to estimate future water production.
- Apply appropriate equations to calculate future performance and plot on Fig. P7.3.5.
- What are the results of your analysis?

Segment	Time (months)	Field History	N_p (MMBO)	GOR (Mcf/BO)	WOR (BW/BO)

Table P7.3.1—Segments encompassing ME 232 well performance.

Generate Fig. P7.3.3 by plotting the last segment on the expanded time scale.
Oil Decline Rate-
WOR Incline Rate-
GOR has declined from 600 scf/BO to approximately 250 scf/BO

State your conclusions.

Nomenclature

a = loss ratio, q/(change in rate)
A = drainage area, ft² (m²)
A_D = dimensionless drainage area
AR = rectangular-drainage-area aspect ratio
b = Arps decline curve exponent
b_f = fracture width, ft (m)
B = fluid formation volume factor, RB/STB (res m³/std m³)
BCPD = stock-tank barrels condensater/day, STB/D (std m³/d)
BOPD = stock-tank barrels oil/day, STB/D (std m³/d)
b_f = fracture width, ft (m)
B_g = gas formation volume factor, res ft³/scf (res m³/std m³)
B_o = oil formation volume factor, RB/STB (res m³/std m³)
b_{pss} = pseudosteady-state intercept
CHP = casinghead pressure, psi (Pa)
C_A = Dietz steady-state shape factor
C_{fD} = dimensionless fracture conductivity
c_f = formation pore compressibility, 1/psi (1/Pa)
c_g = gas compressibility, 1/psi (1/Pa)
c_o = oil compressibility, 1/psi (1/Pa)
C_r = relative conductivity
c_t = total system compressibility, 1/psi (1/Pa)
c_w = water compressibility, 1/psi (1/Pa)
d = effective decline rate, 1/ unit time
D = continuous decline rate, 1/unit of time
D_i = Arps initial decline rate, 1/unit of time
D_{wb} = wellbore diameter, ft (m)
EUR = estimated ultimate recovery, STB (std m³)
FTHP = flowing tubinghead pressure, psi (Pa)
f_w = fraction of water flowing, also called water cut
GOR = gas/oil ratio, Mscf/STB (1000 m³/m³)
GWR = gas/water ratio, Mscf/STBW (1000 std m³/std m³)
G_a = approximate original gas in place estimate, MMscf (million std m³)
G_p = cumulative gas production, MMscf (std m³)
G_{pmax} = estimated ultimate gas cumulative production, MMscf (std m³)
h = reservoir net-pay thickness, ft (m)
HCPV = hydrocarbon pore volume, ft³ (m³)
J = productivity index, STB/D/psi (std m³/s/Pa)
k = reservoir effective permeability, md (m²)
k_f = fracture permeability, md (m²)
$k_f b_f$ = fracture conductivity, md-ft (m³)
k_g = formation effective permeability to gas, md (m²)
kh = formation conductivity, md-ft (m³)
k_o = formation effective permeability to oil, md (m²)

k_w	=	formation effective permeability to water, md (m²)
k_x	=	effective permeability in x-direction, md (m²)
k_y	=	effective permeability in y-direction, md (m²)
k_z	=	vertical effective permeability, md (m²)
L_c	=	system characteristic length, ft (m)
L_D	=	dimensionless horizontal wellbore length
L_h	=	effective horizontal well length in the reservoir, ft (m)
m	=	slope of line
N	=	number of fractures
N	=	original oil in place in the reservoir, STB (std m³)
N_p	=	cumulative oil production, STB (std m³)
$OOIP$	=	original oil in place in the reservoir, STB (std m³)
$OGIP$	=	original gas in place in the reservoir, SCF (std m³)
OWC	=	oil water contact
p_e	=	pressure at external boundary, psi (Pa)
p_i	=	initial reservoir pressure, psi (Pa)
p_p	=	real-gas pseudopressure, psia²/cp (Pa/s)
p_r	=	average reservoir pressure, psia (Pa)
p_{sc}	=	standard-condition pressure, psia (Pa)
p_{tf}	=	flowing tubinghead pressure, psi (Pa)
p_{wD}	=	dimensionless wellbore pressure
p_{wf}	=	sandface flowing pressure, psia (Pa)
q	=	well flow rate, STB/unit of time or Mscf/unit of time (std m³/s)
q_2	=	well flow rate at later time t_2, stb/d or scf/d (std m³/s)
q_D	=	dimensionless flow rate
q_{last}	=	last measured rate, STB/unit of time or scf/unit of time (std m³/s)
q_{Dd}	=	dimensionless decline flow rate
q_{Ddi}	=	dimensionless flow rate integral function
q_{Ddid}	=	dimensionless flow rate integral derivative function
q_{EL}	=	economic-limit flow rate, STB/unit of time (std m³/s)
q_g	=	gas-flow rate, Mscf/unit of time (std m³/s)
q_i	=	initial flow rate, STB/unit of time (std m³/s)
q_o	=	oil-flow rate, STB/unit of time (std m³/s)
Q_{max}	=	maximum possible production @$q = 0$; STB or scf (std m³)
Q_p	=	cumulative fluid production, STB or scf (std m³)
Q_{pD}	=	dimensionless cumulative production
Q_{pDd}	=	dimensionless decline cumulative production
q_{wD}	=	dimensionless well flow rate
r_e	=	reservoir effective drainage radius, ft (m)
r_{eD}	=	dimensionless drainage radius
r_w	=	wellbore radius, ft (m)
r_{wa}	=	apparent wellbore radius, ft (m)
r_{wDz}	=	dimensionless wellbore radius
S	=	steady-state skin effect
S_g	=	gas saturation, fraction of pore volume
S_o	=	oil saturation, fraction of pore volume
S_w	=	water saturation, fraction of pore volume
S_{wi}	=	initial water saturation, fraction of pore volume
t	=	time, days (or months) (s)
T	=	reservoir temperature, °F (K)
t_a	=	real-gas pseudo-time function, D-psi/cp
t_D	=	dimensionless time
t_{DA}	=	dimensionless time referenced to drainage area
t_{Dd}	=	dimensionless decline time
t_{Dmb}	=	dimensionless pseudo production time
t_{mb}	=	pseudo production time, days (s)
t_{mbDd}	=	dimensionless decline pseudo production time

t_p	=	pseudo production time, days (s)
t_{pDd}	=	dimensionless decline pseudo production time
t_{pss}	=	start of boundary-dominated flow in real time
T_{sc}	=	standard-condition temperature, °F or °C
V_b	=	bulk volume of the reservoir, acre-ft or RB (res m^3)
V_p	=	pore volume of the reservoir, acre-ft or RB (res m^3)
WOR	=	water/oil ratio, STBW/stbo (std m^3 water/ std m^3 oil
X_e	=	drainage areal extent in the x-direction, ft (m)
X_{eD}	=	dimensionless drainage extent in x-direction
X_f	=	effective fracture half-length, ft (m)
X_{wD}	=	dimensionless wellbore location in x-direction
Y_e	=	drainage areal extent in the y-direction, ft (m)
Yield	=	STB condensate/MMscf, (std m^3/million std m^3)
Y_{eD}	=	dimensionless drainage extent in y-direction
Y_{wD}	=	dimensionless wellbore location in y-direction
z	=	gas-law deviation (supercompressibility) factor
Z_w	=	horizontal well standoff from bottom of reservoir, ft (m)
Z_{wD}	=	dimensionless horizontal well standoff from bottom of pay
α	=	Prats (1961) dimensionless fracture capacity
Δp	=	pressure difference, psi (Pa)
ε	=	imaging function
ϕ	=	effective porosity of reservoir, fraction of pore volume
γ_f	=	gravity of fluid
λ	=	reservoir mobility of a fluid, md/cp (m^2/Pa·s)
λ_t	=	total system mobility (all fluids), md/cp (m^2/ Pa·s)
μ	=	reservoir fluid viscosity, cp (Pa·s)
μ_g	=	gas viscosity, cp (Pa·s)
μ_{gi}	=	initial reservoir gas viscosity, cp (Pa·s)
μ_o	=	oil viscosity, cp (Pa·s)
μ_{oi}	=	initial oil viscosity, cp (Pa·s)

References

Agarwal, R. G., Gardner, D. C., Kleinsteiber, S. W. et al. 1998. Analyzing Well Production Data Using Combined Type Curve and Decline Curve Analysis Concepts. Presented at the SPE Annual Technical Conference and Exhibition, New Orleans, 27–30 September. SPE-49222-MS. https://doi.org/10.2118/49222-MS.

Aguilera, R. and Ng, M. C. 1991. Decline-Curve Analysis of Hydraulically Fractured Wells in Dual-Porosity Reservoirs. Presented at the SPE Annual Technical Conference and Exhibition, Dallas, 6–9 October. SPE-22938-MS. https://doi.org/10.2118/22938-MS.

Anderson, D. M., Nobakht, M., Moghadam, S. et al. 2010. Analysis of Production Data From Fractured Shale Gas Wells. Presented at the SPE Unconventional Gas Conference, Pittsburgh, Pennsylvania, USA, 23–25 February. SPE-131787-MS. https://doi.org/10.2118/131787-MS.

Ansah, J., Knowles, R. S., and Blasingame, T. A. 1996. A Semi-Analytic (p/z) Rate-Time Relation for the Analysis and Prediction of Gas Well Performance. Presented at the SPE Mid-Continent Gas Symposium, Amarillo, Texas, USA, 28–30 April. SPE-35268-MS. https://doi.org/10.2118/35268-MS.

Arps, J. J. 1945. Analysis of Decline Curves. In *Transactions of the Society of Petroleum Engineers*, Vol. 160, Number 1, 228–247. SPE-945228-G. Richardson, Texas: SPE. https://doi.org/10.2118/945228-G.

Azari, M., Soliman, M. Y., Wooden, W. O. et al. 1991. Performance Prediction for Finite-Conductivity Vertical Fractures. Presented at the SPE Annual Technical Conference and Exhibition, Dallas, 6–9 October. SPE-22659-MS. https://doi.org/10.2118/22659-MS.

Baker, R. O., Anderson, T., and Sandhu, K. 2003. Using Decline Curves to Forecast Waterflooded Reservoirs: Fundamentals and Field Cases. Presented at the Canadian International Petroleum Conference, Calgary, 10–12 June. PETSOC-2003-181. https://doi.org/10.2118/2003-181.

Berg, R. R. 1986. *Reservoir Sandstones.* New York: Prentice-Hall.

Blasingame, T. A. and Lee, W. J. 1986. Variable-Rate Reservoir Limits Testing. Presented at the Permian Basin Oil and Gas Recovery Conference, Midland, Texas, USA, 13–14 March. SPE-15028-MS. https://doi.org/10.2118/15028-MS.

Blasingame, T. A., McCray, T. L., and Lee, W. J. 1991. Decline Curve Analysis for Variable Pressure Drop/Variable Flowrate Systems. Presented at the SPE Gas Technology Symposium, Houston, 22–24 January. SPE-21513-MS. https://doi.org/10.2118/21513-MS.

Bondar, V. V. and Blasingame, T.A. 2002. Analysis and Interpretation of Water-Oil-Ratio Performance. Prepared for presentation at the SPE Annual Technical Conference and Exhibition, San Antonio, Texas, USA, 29 September–2 October.

Boyer, C., Kieschnick, J., Lewis, R. E. et al. 2006. Producing Gas From Its Source. *Oilfield Review* **18** (3): 36–49.

Brons, F. 1963. On the Use and Misuse of Production Decline Curves. *Prod. Monthly* **27** (9): 22–25.

Buba, I. M. 2003. *Direct Estimation of Gas Reserves Using Production Data*. MSc thesis, Texas A&M University, College Station, Texas (August 2003).

Buckley, S. E. and Leverett, M. C. 1942. Mechanism of Flood Displacement in Sands. In *Transactions of the Society of Petroleum Engineers,* Vol. 146, Number 1, 107–116, SPE-942107-G. Richardson, Texas: SPE. https://doi.org/10.2118/942107-G.

Chan, K. S. 1995. Water Control Diagnostic Plots. Presented at the SPE Annual Technical Conference and Exhibition, Dallas, 22–25 October. SPE-30775-MS. https://doi.org/10.2118/30775-MS.

Cinco-Ley, H. and Meng, H. Z. 1988. Pressure Transient Analysis of Wells With Finite Conductivity Vertical Fractures in Double Porosity Reservoirs. Presented at the SPE Annual Technical Conference and Exhibition, Houston, 2–5 October. SPE-18172-MS. https://doi.org/10.2118/18172-MS.

Chen, C. and Raghavan, R. 1997. A Multiply-Fractured Horizontal Well in a Rectangular Drainage Region. *SPE J.* **2** (4): 455–465. SPE-37072-PA. https://doi.org/10.2118/37072-PA.

Chen, C. and Raghavan, R. 2013. On Some Characteristic Features of Fractured-Horizontal Wells and Conclusions Drawn Thereof. *SPE Res Eval & Eng* **16** (1): 19–28. SPE-163104-PA. https://doi.org/10.2118/163104-PA.

Chu, W., Fleming, C. H., and Carroll, K. M. 2001. Determination of Original Gas in Place in Bollycotton, Offshore Ireland. *SPE Res Eval & Eng* **4** (1): 11–15. SPE-69735-PA. https://doi.org/10.2118/69735-PA.

Cinco-Ley, H., Samaniego-V., F., and Dominguez-A., N. 1978. Transient Pressure Behavior for a Well With a Finite-Conductivity Vertical Fracture. *SPE J.* **18** (4): 253–264. SPE-6014-PA. https://doi.org/10.2118/6014-PA.

Cobb, W. M. and Marek, F. J. 1997. Determination of Volumetric Sweep Efficiency in Mature Waterfloods Using Production Data. Presented at the SPE Annual Technical Conference and Exhibition, San Antonio, Texas, USA, 5–8 October. SPE-38902-MS. https://doi.org/10.2118/38902-MS.

Cutler, W. W. Jr. 1924. Estimation of Underground Oil Reserves by Oil-Well Production Curves. Bulletin 228, Department of the Interior, Bureau of Mines, Washington, DC (August 1924).

Dembicki, H. 2009. Three Common Source Rock Evaluation Errors Made by Geologists During Prospect or Play Appraisals. *AAPG Bull.* **93** (3): 341–356. https://doi.org/10.1306/10230808076.

Dietz, D. N. 1965. Determination of Average Reservoir Pressure From Build-Up Surveys. *J Pet Technol* **17** (8): 955–959. SPE-1156-PA. https://doi.org/10.2118/1156-PA.

Dou, H., Chen, C., Change, Y. W. et al. 2007. Decline Analysis for Horizontal Wells of Intercampo Field, Venezuela. Presented at the SPE Production and Operations Symposium, Oklahoma City, Oklahoma, USA, 31 March–3 April. SPE-106440-MS. https://doi.org/10.2118/106440-MS.

Doublet, L. E. and Blasingame, T. A. 1996. Evaluation of Injection Well Performance Using Decline Type Curves. Paper prepared for presentation at the SPE Permian Basin Oil and Gas Recovery Conference, Midland, Texas, USA, 27–29 March.

Doublet, L. E. and Blasingame, T. A. 1995. Decline Curve Analysis Using Type Curves: Water Influx/Waterflood Cases. Paper prepared for presentation at the SPE Annual Technical Conference and Exhibition, Dallas, 22–25 October.

Economides, M. J. and Nolte, K. G. 2000. *Reservoir Stimulation*, 3rd edition. Chichester, England: John Wiley & Sons Ltd.

England, K. W., Poe, B. D., and Conger, J. G. 2000. Comprehensive Evaluation of Fractured Gas Wells Utilizing Production Data. Presented at the SPE Rocky Mountain Regional/Low-Permeability Reservoirs Symposium and Exhibition, Denver, 12–15 March. SPE-60285-MS. https://doi.org/10.2118/60285-MS.

Equity SubCommittee, Long Beach Unit. 1964. The Method for Establishing Recovery Factors for the Long Beach Unit. City of Long Beach (23 September 1964).

Ershaghi, I. and Abdassah, D. 1984. A Prediction Technique for Immiscible Processes Using Field Performance Data (includes associated papers 13392, 13792, 15146, and 19506). *J Pet Technol* **36** (4): 664–670. SPE-10068-PA. https://doi.org/10.2118/10068-PA.

Fekete. 2012. Fast Well Test Analysis. Fekete Assoc. Inc

Fetkovich, M. J. 1980. Decline Curve Analysis Using Type Curves. *J Pet Technol* **32** (6): 1065–1077. SPE-4629-PA. https://doi.org/10.2118/4629-PA.

Fetkovich, M. J. and Thrasher, T. S. 1979. Constant Well Pressure Testing and Analysis in Low Permeability Reservoirs. Paper prepared for presentation at the SPE/DOE Symposium on Low Permeability Gas Reservoirs, Denver, 20–22 May.

Fetkovich, M. J., Vienot, M. E., Bradley, M. D. et al. 1987. Decline Curve Analysis Using Type Curves: Case Histories. *SPE Form Eval* **2** (4): 637–656. SPE-13169-PA. https://doi.org/10.2118/13169-PA.

Fraim, M. I. and Wattenbarger, R. A. 1987. Gas Reservoir Decline-Curve Analysis Using Type Curves With Real Gas Pseudopressure and Normalized Time. *SPE Form Eval* **2** (4): 671–682. SPE-14238-PA. https://doi.org/10.2118/14238-PA.

Gringarten, A. C., Ramey, H. J. Jr., and Raghavan, R. 1974. Unsteady-State Pressure Distributions Created by a Well With a Single Infinite-Conductivity Vertical Fracture. *SPE J.* **14** (4): 347–360. SPE-4051-PA. https://doi.org/10.2118/4051-PA.

Guimeráns, R, Tovar, J., Pinto, F. et al. 1997. Application of New Technologies to Horizontal Wells in Venezuela. Presented at the Latin American and Caribbean Petroleum Engineering Conference, Rio de Janeiro, Brazil, 30 August–3 September. SPE-39072-MS. https://doi.org/10.2118/39072-MS.

Guppy, K. H., Cinco-Ley, H., and Ramey, H. J. Jr. 1981. Transient Flow Behavior of a Vertically Fractured Well Producing at Constant Pressure. SPE-9963-MS, eLibrary, unsolicited.

Holditch, S. A. 2014. Class Notes, Texas A&M University, College Station, Texas.

Horner, D. R. 1951. Pressure Build-Up in Wells. *Proc.*, Third World Petroleum Congress, The Hague, The Netherlands, II (1951), 503.

Huddleston, B. P. 1991. Class Notes, Texas A&M University, College Station, Texas.

Hutton, A., Bharati, S., and Robl, T. 1994. Chemical and Petrographic Classification of Kerogen/Macerals. *Energy Fuels* **8** (6): 1478–1488. https://doi.org/10.1021/ef00048a038.

Jordan, C. L., Smith, C. R, and Jackson, R. A. 2009. A Rapid and Efficient Production Analysis Method for Unconventional and Conventional Gas Reserves. Presented at the Asia Pacific Oil and Gas Conference and Exhibition, Jakarta, Indonesia, 4–6 August. SPE-120737-MS. https://doi.org/10.2118/120737-MS.

Knowles, S. K. 1996. *Development and Verification of Analysis Relations for the Surface Testing of Gas Wells*. MSc thesis, Texas A&M University, College Station, Texas.

Langmuir, I. 1918. The Adsorption of Gases on Plane Surfaces of Glass, Mica and Platinum. *J. Am. Chem. Soc.* **40** (9): 1361–1403. https://doi.org/10.1021/ja02242a004.

Lijek, S. J. 1989. Simple Performance Plots Used in Rate-Time Determination and Waterflood Analysis. Presented at the SPE Annual Technical Conference and Exhibition, San Antonio, Texas, USA, 8–11 October. SPE-19847-MS. https://doi.org/10.2118/19847-MS.

Maley, S. 1985. The Use of Conventional Decline Curve Analysis in Tight Gas Well Applications. Presented at the SPE/DOE Low Permeability Gas Reservoirs Symposium, Denver, 19–22 March. SPE-13898-MS. https://doi.org/10.2118/13898-MS.

Matthews, C. S. and Russell, D. G. 1967. *Pressure Buildup and Flow Tests in Wells*. Richardson, Texas: Society of Petroleum Engineers.

McCray, T. L. 1990. *Reservoir Analysis Using Production Data and Adjusted Time*. MSc thesis, Texas A&M University, College Station, Texas.

Miller, H. C. 1942. Oil Reservoir Behavior Based on Pressure–Production Data. US Bureau of Mines R.I.3634, US Department of the Interior, Washington, DC, 36 pages.

Newsham, K. E. and Rushing, J. A. 2001. An Integrated Work-Flow Model to Characterize Unconventional Gas Reservoirs: Part 1—Geological Assessment and Petrophysical Evaluations. Presented at the SPE Annual Technical Conference and Exhibition, New Orleans, 30 September–3 October. SPE-71351-MS. https://doi.org/10.2118/71351-MS.

Ozkan, E. 1988. *Performance of Horizontal Wells*. PhD dissertation, University of Tulsa, Oklahoma.

Ozkan, E., Brown, M. L., Raghavan, R. et al. 2011. Comparison of Fractured-Horizontal-Well Performance in Tight Sand and Shale Reservoirs. *SPE Res Eval & Eng* **14** (2): 248–259. SPE-121290-PA. https://doi.org/10.2118/121290-PA.

Ozkan, E. and Raghavan, R. 1991. New Solutions for Well-Test-Analysis Problems: Part 1—Analytical Considerations (includes associated papers 28666 and 29213). *SPE Form Eval* **6** (3): 359–368. SPE-18615-PA. https://doi.org/10.2118/18615-PA.

Palacio, J. C. and Blasingame, T. A. 1993. Decline-Curve Analysis Using Type Curves—Analysis of Gas Well Production Data. Oral presentation at the Low Permeability Reservoirs Symposium, Denver, 26–28 April. SPE-25909-MS.

Poe, B. D. Jr. 2003. Production Diagnostic Analyses With Incomplete or No Pressure Records. Presented at the SPE Annual Technical Conference and Exhibition, Denver, 5–8 October. SPE-84224-MS. https://doi.org/10.2118/84224-MS.

Poe, B. D. Jr. and Poston, S. W. 2010. Evaluation of Time to Flow Stabilization and Effective Drainage Area of Slow-Stabilizing Wells Using Production Decline Analysis. Presented at the SPE Annual Technical Conference and Exhibition, Florence, Italy, 19–22 September. SPE-135475-MS. https://doi.org/10.2118/135475-MS.

Poe, B. D. Jr. and Marhaendrajana, T. 2002. Investigation of the Relationship Between the Dimensionless and Dimensional Analytic Transient Well Performance Solutions in Low-Permeability Gas Reservoirs. Presented at the SPE Annual Technical Conference and Exhibition, San Antonio, Texas, USA, 29 September–2 October. SPE-77467-MS. https://doi.org/10.2118/77467-MS.

Poston, S. W. and Poe, B. D. Jr. 2008. *Analysis of Production Decline Curves*. Richardson, Texas: Society of Petroleum Engineers.

Prats, M. 1961. Effect of Vertical Fractures on Reservoir Behavior—Incompressible Fluid Case. *SPE J.* **1** (2): 105–118. SPE-1575-G. https://doi.org/10.2118/1575-G.

Raghavan, R. and Joshi, S. D. 1993. Productivity of Multiple Drainholes or Fractured Horizontal Wells. *SPE Form Eval* **8** (1): 11–16. SPE-21263-PA. https://doi.org/10.2118/21263-PA.

Ramey, H. J. Jr. and Cobb, W. M. 1971. A General Pressure Buildup Theory for a Well in a Closed Drainage Area (includes associated paper 6563). *J Pet Technol* **23** (12): 1493–1505. SPE-3012-PA. https://doi.org/10.2118/3012-PA.

Seewald, J. S. 2003. Organic–Inorganic Interaction in Petroleum-Producing Sedimentary Basins. *Nature* **426** (November): 327–333. https://doi.org/10.1038/nature02132.

Shih, M. Y. and Blasingame, T. A. 1995. Decline Curve Analysis Using Type Curves. Paper prepared for presentation at the SPE Rocky Mountain Low Permeability Meeting, Denver, 19–22 March.

Stiles, W. E. 1949. Use of Permeability Distribution in Water Flood Calculations. *J Pet Technol* **1** (1): 9–13. SPE-949009-G. https://doi.org/10.2118/949009-G.

Timmerman, E. H. 1971. Predict Performance of Water Floods Graphically. *Pet. Eng.* **43** (12).

van Everdingen, A. F. and Hurst, W. 1949. The Application of the Laplace Transformation to Flow Problems in Reservoirs. *J Pet Technol* **1** (12): 305–324. SPE-949305-G. https://doi.org/10.2118/949305-G.

van Krevelen, D. W. 1950. Graphical-Statistical Method for the Study of Structure and Reaction Processes of Coal. *Fuel* **29**: 269–284.

Wattenbarger, R. A., El-Banbi, A. H., Villegas, M. E. et al. 1998. Production Analysis of Linear Flow Into Fractured Tight Gas Wells. Presented at the SPE Rocky Mountain Regional/Low-Permeability Reservoirs Symposium, Denver, 5–8 April. SPE-39931-MS. https://doi.org/10.2118/39931-MS.

Welge, H. J. 1952. A Simplified Method for Computing Oil Recovery by Gas or Water Drive. *J Pet Technol* **4** (4): 91–98. SPE-124-G. https://doi.org/10.2118/124-G.

Willis, M. and Tutuncu, A. N. 2014. Integration of Core, Drilling, Microseismic and Well Log Data for Geomechanical Property Determination and Monitoring in the Argentinian Vaca Muerta Shale Formation. Presented at the SPE/AAPG/SEG Unconventional Resources Technology Conference, Denver, 25–27 August. URTEC-1922481-MS. https://doi.org/10.15530-URTEC-2014-1922481.

Zerpa, L. B. 1995. Numerical Simulation of Horizontal Wells in a Heavy Crude Reservoir in Venezuela. Presented at the SPE International Heavy Oil Symposium, Calgary, 19–21 June. SPE-30282-MS. https://doi.org/10.2118/30282-MS.

INDEX

A

Arps equations
    bounds of, 11
    degree of curvature, 5
    exponential curve
        Arps nominal decline, 7
        b-exponent term, 6
        Cartesian rate *vs.* cumulative production plots, 9
        constant and continuous declines, 7–8
        constant percentage exponential decline, 6–7
        EUR, 7
        exponential rate *vs.* time expression, 6
        hyperbolic rate-time expression, 6
        rate *vs.* cumulative production plot, 8–9
        rate *vs.* time plot, 8, 9
        semilog rate *vs.* time, 9
    harmonic equations, 10–11
    hyperbolic equations, 9–10
    initial decline rate, 5
    initial producing rate, 5
    loss ratio, 5
Arps (1945) late-time production decline model behavior, 89–91

B

Blasingame et al. method
    assumptions and characteristics, 97–98
    composite type curve, 98
    correlating functions, 98–101
    decline curve solution, 102
    empirical scaling term, 102
    horizontal well decline curves, 102
    integral and integral-derivative function transformations, 85
    matching procedure, 101–102
    normalized rate and pressure changes, 97
    production rate normalization, 97
    step- and ramp-rate boundary flux models, 102
Bollycotton Gas Field
    Cartesian rate *vs.* cumulative production, 40, 42
    G_a and OGIP plot, 40
    initial condition (p/z) relationship, 39
    logarithmic rate *vs.* time plot, 40, 41
    multiplot analysis, 40
    original gas in place calculation, 39
    performance history, 28–29, 40, 41
    producing and reservoir properties, 42
    quadratic solution, 40, 42
    in rate *vs.* time plot, 39
    reservoir properties and OGIP, 39
    semilog rate *vs.* time, 40, 41
    straight line approximation, 40
boundary-dominated flow segment (BDF), 38
Buckley–Leverett plot, 126–127

C

carbon/hydrogen (C/H) ratio, 69
character production curve, 2

D

Darcy equation, 49, 64, 65
decline curves
    advanced decline curve analysis, 2
    applications, 2
    Arps equations (*see* Arps equations)
    assumptions, 4
    boundary-dominated flow, 3–4
    character production curve, 2
    drainage volume, 2
    of dually completed well, 16–17
    expanding drainage radius, 3
    of Hollands No. 3A well, 14–15
    of north Texas gas condensate well, 17–18
    oil and gas production rates decline, 2
    production rates, 2
    reservoir drainage limits, 3
    shapes of production decline curves, 4, 5
    transient and boundary-dominated flow periods, 2
    transient boundary-dominated conditions
        Glenn Pool Field, depletion history, 12–14
        log rate *vs.* log time plot, 12
        North Sea field, depletion history, 12
    unforeseen water production, performance analysis, 17
    very-low-permeability gas case, 4
    of Wafford No. 1 well, rate *vs.* time history, 14
depletion model
    b-exponent, 73–74
    conventional plotting methods, 74
    drainage area aspect ratio, 71
    flow system, 72
    fracture stages, 71
    normalizing curves, 74
    pressure-transient solution, 72
    pseudo-linear flow behavior, 72
    reciprocal flow rate *vs.* square root of time, 72
    reciprocal rate-transient solution, 72
    stimulated reservoir volume, 71, 72
    system characteristic length, 72
diffusivity equation
    application, 48
    constant pressure solution, 51
    constant rate solution, 50–51
    definition, 49
    van Everdingen and Hurst solutions, 49–50
Dirichlet inner boundary condition, 86
drainage area aspect ratio (AR), 71
dual-porosity Austin Chalk well, production history, 20, 21

E

effects of field conditions
    Bollycotton Gas Field, performance history, 28–29
    different-scale performance plots, 19
    dual-porosity Austin Chalk well, production history, 20, 21
    Ellenberger gas well, performance history, 20, 28
    Gulf of Mexico field, performance history, 26–27
    informational plots
        API gravity, 26
        Salt Creek Field, production history, 24, 25
        well records, 25
    multiplot analysis, 26
    production segments
        L95 well, production history, 23
        production history, matching curves, 22
        waterflooded Canadian reservoir, 21–22
    production system schematic diagram, 19, 20
    water influx effect, 20, 21
    well downtime, 23, 24
    west Texas hydraulically fractured oil well, performance history, 27
Ellenberger gas well, 20
Ershagi plot, 136, 140
estimated ultimate recovery (EUR), 7, 10, 15, 37, 74, 134, 141

F

Fetkovich method
    Arps (1945) late-time production decline model behavior, 89–91
    boundary-dominated flow stems, 87
    cumulative production, 88
    decline analysis time, 88
    decline curve model, 89
    dimensionless decline flow rate, 87, 88
    dimensionless decline flow time, 87, 88
    dimensionless drainage radius, 88, 89
    flow rate variables, 88, 89
    Golden Zuma well, productive character, 109–112
    graphical scaling parameters, 85
    infinite-acting transient, 87
    logarithmic rate *vs.* time curve, 108, 109
    logarithmic transformations, 88
    rate-transient dimensionless time, 89
    semilog rate *vs.* time plot, 108, 109
    transient and bounded flow production history, 85
    transient decline curve
        analysis procedure, 93
        apparent wellbore radius and skin, 92, 96
        boundary-dominated example, 93–94
        bulk volume of reservoir, 96
        computer-generated type curve match, 95, 96
        drainage area, 92
        drainage volume, 96
        formation conductivity, 92
        infinite-acting transient flow behavior, 95
        log rate *vs.* log time plot, 94, 95
        matching of production data, 91
        M– 4X well production history, 94, 95
        permeability calculation, 95
        rate and time solutions, 91–92
        rate-transient performance, 91
        reservoir drainage area, 97
        reservoir information, 92
        reservoir pore volume, 92
        skin factor, 96
        total compressibility calculation, 96
        total system compressibility, 92
    transient flow studies, 89
    transient side analysis, 97
    unique decline curve analysis, 88
flowing bottomhole pressures (FBHP), 74
fractured horizontal wells
    depletion model
        b-exponent, 73–74
        conventional plotting methods, 74
        drainage area aspect ratio, 71
        flow system, 72
        fracture stages, 71
        normalizing curves, 74
        pressure-transient solution, 72
        pseudo-linear flow behavior, 72
        reciprocal flow rate *vs.* square root of time, 72
        reciprocal rate-transient solution, 72
        stimulated reservoir volume, 71, 72
        system characteristic length, 72
    early time behavior, 82–84
    geological setting
        capillary effects, 67
        carbonate sediments, 68
        clay- and silt-sized sediments, 68
        current-velocity and grain-size effects, 68
        hydrocarbon generation, 69–70
        maturation and generation of hydrogen, 70–71
        oil- and gas-saturated shales, 67
        potential problems, 68
        productive shales, 67
    regional analysis example
        interpretation, 79, 81
        log production rate *vs.* log time analysis, 79, 80
        plotting methods, 78
        reciprocal flow rate *vs.* square root of time analysis, 79, 80
        south Texas Eagle Ford Shale lease, horizontal well completions, 79
        time ratio plot, 79, 80
        well production data, 79
    shale well examples
        Fetkovich type curve, 75, 78
        Hixon oil well production history, 75
        linear flow parameter, 77
        matrix/fracture interface area, 77
        normalized flow rate and square root of pseudo production time smoothed data, 75, 77
        pressure drop normalized flow rate *vs.* pseudo production time plot, 75, 76
        pressure drop normalized production flow rate *vs.* time plot, 75, 76
        $1/q$ *vs.* $\sqrt{t}$ plot, 78
        reciprocal flow rate *vs.* square root of time, 75, 77
        t_{mb}/t ratio, 75, 76
fractured vertical wellbore case
    b-exponent, 58–59
    dimensionless fracture conductivity, 55–57
    finite conductivity, 55
    formation damage, 61, 62
    fracture conductivity, 55
    geometrical relationship, 55
    infinite-conductivity-vertical-fracture response, 55
    log flow rate *vs.* log time plot, 59
    performance history, 57, 58
    principal transient flow regimes, 57
    production performance, 55
    reciprocal flow rate *vs.* square root of time plot, 60
    straight-line extrapolation, 61, 62
frontal advance theory, 128–130

G

Glenn Pool Field
  depletion history, 12, 13
  initial production decline curve, 13
  performance histories, 13, 14
  production history, 12, 13
  reserves to production ratio, 14
Golden Zuma well, productive character, 109–112
Gulf of Mexico (GOM) field, performance history, 26–27

H

Hixon oil well production history, 75
horizontal-unfractured-well case, 51–52
hydrogen/carbon (H/C) ratio, 69

K

Kentucky well, 113–114

M

ME 232 well, production history, 149, 151–153
multiple performance plots, two phase flow
  Cartesian rate *vs.* time plot, 136, 138
  Cartesian scale WOR *vs.* time plot, 136, 139
  composite performance history, 136, 137
  decline rate, 141
  Ershagi plot, 136, 140
  fractional flow and WOR results, 141
  oil and water production histories, 136, 139
  oil rate *vs.* cumulative recovery plot, 136, 138
  relating cumulative production to producing time plot, 136
  semi-log rate *vs.* time plot, 136, 138
  total well liquid flow rate, 136, 139
  watercut plot, 136, 140
  well history, 136
  1/WOR + WOR plot, 136, 140

N

Neumann inner boundary condition, 86
normal decline curve analysis, 115
North Sea field, depletion history, 12

O

oil/water contact (OWC), 116
oxygen/carbon (O/C) index ratio, 69

P

Poe and Poston method
  advantages, 104
  completion models
    composite decline curves, 107
    finite conductivity vertically fractured well, 106
    horizontal wellbore case, 107–108
    unfractured vertical well, 105
    vertical fracture decline curves, 107
  composite decline curves, 103
  computer-aided analysis, 102
  dimensionless cumulative production, 103
  dimensionless decline cumulative production, 104
  dimensionless decline flow rate, 103
  dimensionless pseudo production time, 103
  graphical analysis technique, 85
  imaging factor, 103
  iterative computations, 104
  pressure drop normalized cumulative production, 104
  rate equations, 104
  reservoir parameters, 103
  vertical well decline curve match and analysis, 103

Provost Field, performance history
  Arps curve fits, 122, 123
  GOR and water-cut performance history, 121, 122
  oil, total liquid producing rate, and water injection history, 121, 122
  segments of, 122, 123
  water handling costs, 123, 124
  well history water handling costs, 121, 122
pseudo production time
  characteristic length, 32
  definition, 31
  dimensionless material balance time, 32
  equivalent material balance time, 31, 32
  superposition dimensionless time
    fractured vertical well, 32–33
    horizontal wellbore length, 33–34
    material balance, 32
    unfractured well, 32, 33

R

regional analysis example
  interpretation, 79, 81
  log production rate *vs.* log time analysis, 79, 80
  plotting methods, 78
  reciprocal flow rate *vs.* square root of time analysis, 79, 80
  south Texas Eagle Ford Shale lease, horizontal well completions, 79
  time ratio plot, 79, 80
  well production data, 79

S

Salt Creek Field, production history, 24, 25
shale well examples
  Fetkovich type curve, 75, 78
  Hixon oil well production history, 75
  linear flow parameter, 77
  matrix/fracture interface area, 77
  normalized flow rate and square root of pseudo production time smoothed data, 75, 77
  pressure drop normalized flow rate *vs.* pseudo production time plot, 75, 76
  pressure drop normalized production flow rate *vs.* time plot, 75, 76
  $1/q$ *vs.* $\sqrt{t}$ plot, 78
  reciprocal flow rate *vs.* square root of time, 75, 77
  t_{mb}/t ratio, 75, 76
smoothing variable production
  assumptions and limitations, 37
  Bollycotton Gas Field
    Cartesian rate *vs.* cumulative production, 40, 42
    G_a and OGIP plot, 40
    initial condition (p/z) relationship, 39
    logarithmic rate *vs.* time plot, 40, 41
    multiplot analysis, 40
    original gas in place calculation, 39
    performance history, 40, 41
    producing and reservoir properties, 42
    quadratic solution, 40, 42
    in rate *vs.* time plot, 39
    reservoir properties and OGIP, 39
    semilog rate *vs.* time, 40, 41
    straight line approximation, 40
  boundary-dominated flow segment, 38
  computer programs, 31
  exponential and quadratic curves, 36, 37
  flowing pressure history, 37
  initial pressure, 37
  initial producing rate, 37

normalizing production, 34–35
pseudo production time
  characteristic length, 32
  definition, 31
  dimensionless material balance time, 32
  equivalent material balance time, 31, 32
  superposition dimensionless time, 32–34
quadratic equation, 31, 37, 38
quadratic model, 35–36
West Virginia Well A analysis problem
  Cartesian rate vs. cumulative production, 42, 43
  flow rate cumulative production plot, 45
  logarithmic rate-logarithmic time plot, 44
  logarithmic rate vs. log time, 42, 43
  log flow rate vs. time plot, 45
  log rate vs. time, 42, 43
  quadratic equation, 42, 44
  quadratic plot, 45
  summary analysis, 44
z-factor, 37
stimulated reservoir volume (SRV), 71, 72

T
two phase flow
  geological influences, 116
  ME 232 well, production history, 149, 151–153
  multiple performance plots
    Cartesian rate vs. time plot, 136, 138
    Cartesian scale WOR vs. time plot, 136, 139
    composite performance history, 136, 137
    decline rate, 141
    Ershagi plot, 136, 140
    fractional flow and WOR results, 141
    oil and water production histories, 136, 139
    oil rate vs. cumulative recovery plot, 136, 138
    relating cumulative production to producing time plot, 136
    semi-log rate vs. time plot, 136, 138
    total well liquid flow rate, 136, 139
    watercut plot, 136, 140
    well history, 136
    1/WOR + WOR plot, 136, 140
  performance history, Provost Field
    Arps curve fits, 122, 123
    GOR and water-cut performance history, 121, 122
    oil, total liquid producing rate, and water injection history, 121, 122
    segments of, 122, 123
    water handling costs, 123, 124
    well history water handling costs, 121, 122
  relative permeability
    Buckley–Leverett plot, 131, 133
    definition, 127
    field parameters, 131
    fractional flow curves, 134
    frontal advance theory, 128–130
    GOR and drilling history, 131, 132
    historical time and cumulative production, 131, 133
    interpretation of field history, 132
    interpretive concepts, 134–136
    k_w/k_o ratio curve, 127
    oil and water relative permeability curves, 127, 128
    oil production rate and WOR, 131, 132
    primary and subordinate stages, 133–134
    pseudorelative permeability, 130–131
    segment parameters, 132, 133
    straight-line approximation, 131
  reserves and predicting performance
    analysis procedure, 124–126
    Arps's equations, 123
    assumptions, 124
    Buckley–Leverett plot, 126–127
    cumulative production, 124
    hyperbolic equations, 124
    oil flow rate and WOR predictions, 126, 127
    rate equation, 124
    solution procedure, 124
  waterflooding
    capillary force, 118
    capillary/viscous number, 118
    constant vigilance, 115
    degree of crossflow, 117
    depletion stages, 120
    efficiency, 115
    flood-front pattern, 117
    gravity force, 118
    incremental production, 115
    linearization technique, 120
    material balance techniques, 117
    permeability distribution, 117
    phase flow relationships, 115
    pseudorelative permeability curves, 119
    two-step process, 117
    viscous force, 118
    viscous/gravity number, 118
    water- to oil-mobility ratio, 117
    WOR vs. N_p plot, 119
  well diagnostics plots
    Cartesian rate vs. time, 146, 147
    derivative plots, 143–144
    Evans Well 11, 146, 147
    GOR and derivative history, 144, 145
    log rate vs. time, 146, 148
    near-wellbore channeling, 142
    oil and water producing rates, 146, 148
    performance history, 142
    producing history, 144, 145
    rate vs. cumulative production, 146, 148
    total system flow rate, 146, 149
    water cones, 142
    Well MR-321, 144–146, 149–151
    WOR vs. time plot, 142
  Well PP4, production history, 147, 149
type curves
  apparent wellbore radius calculation, 112
  Blasingame et al. method
    assumptions and characteristics, 97–98
    composite type curve, 98
    correlating functions, 98–101
    decline curve solution, 102
    empirical scaling term, 102
    horizontal well decline curves, 102
    integral and integral-derivative function transformations, 85
    matching procedure, 101–102
    normalized rate and pressure changes, 97
    production rate normalization, 97
    step- and ramp-rate boundary flux models, 102
    water influx and waterflood performance, 102
  boundary-dominated flow regime, 86
  definition, 86
  dimensionless and field data match points, 86
  dimensionless time, 87
  dimensionless well flow rate solution, 87
  Dirichlet inner boundary condition, 86
  Fetkovich method
    Arps (1945) late-time production decline model behavior, 89–91
    boundary-dominated flow stems, 87

cumulative production, 88
decline analysis time, 88
decline curve model, 89
dimensionless decline flow rate, 87, 88
dimensionless decline flow time, 87, 88
dimensionless drainage radius, 88, 89
flow rate variables, 88, 89
graphical scaling parameters, 85
infinite-acting transient, 87
logarithmic transformations, 88
rate-transient dimensionless time, 89
transient and bounded flow production history, 85
transient decline curve, 91–97
transient flow studies, 89
unique decline curve analysis, 88
0 to 1 flow regime, 86
1 to 2 flow regime, 86
2 to 3 flow regime, 86
formation permeability calculation, 112
general production rate decline flow regimes, 86
inner boundary condition, 86
Kentucky well, 113–114
Neumann inner boundary condition, 86
Poe and Poston method
    advantages, 104
    completion models, 105–108
    composite decline curves, 103
    computer-aided analysis, 102
    dimensionless cumulative production, 103
    dimensionless decline cumulative production, 104
    dimensionless decline flow rate, 103
    dimensionless pseudo production time, 103
    graphical analysis technique, 85
    imaging factor, 103
    iterative computations, 104
    pressure drop normalized cumulative production, 104
    rate equations, 104
    reservoir parameters, 103
    vertical well decline curve match and analysis, 103
quadratic equation, 113
reference type curve solutions, 86
reservoir drainage pore volume calculation, 113
skin effect evaluation, 112
system characteristic length, 87
terminal pressure inner boundary condition, 86–87
transition flow period, 86

U
unfractured vertical wellbore models, 48–49

V
van Krevelen plot, 69

W
waterdrive displacement mechanisms, 116
well and reservoir models
  analysis procedure
    Cartesian flow rate *vs.* cumulative recovery plot, 61
    log flow rate *vs.* log time plot, 61
    production data, 61
    $1/q$ *vs.* $\sqrt{t}$ plot, 61, 63, 64
    semilog flow rate *vs.* time plot, 61, 63, 64
    time ratio plot, 63
  flow equations
    Darcy equation, 49, 64, 65
    diffusivity equation, 48–51

fractured vertical wellbore case
    b-exponent, 58–59
    dimensionless fracture conductivity, 55–57
    finite conductivity, 55
    formation damage, 61, 62
    fracture conductivity, 55
    geometrical relationship, 55
    infinite-conductivity-vertical-fracture response, 55
    log flow rate *vs.* log time plot, 59
    performance history, 57, 58
    principal transient flow regimes, 57
    production performance, 55
    reciprocal flow rate *vs.* square root of time plot, 60
    straight-line extrapolation, 61, 62
geological considerations, 48
horizontal-unfractured-well case, 51–52
horizontal well
    learning objectives, 52
    performance history, 53–54
    transient production performance, 52
    turbidities, 52
producing interval transmissibility, 47
reservoir geologic factors, 47
unfractured vertical wellbore models, 48–49
well completion configurations, 47
well stimulation techniques, 47
well diagnostics plots
    Cartesian rate *vs.* time, 146, 147
    derivative plots, 143–144
    Evans Well 11, 146, 147
    GOR and derivative history, 144, 145
    log rate *vs.* time, 146, 148
    near-wellbore channeling, 142
    oil and water producing rates, 146, 148
    performance history, 142
    producing history, 144, 145
    rate *vs.* cumulative production, 146, 148
    total system flow rate, 146, 149
    water cones, 142
    Well MR-321, 144–146, 149–151
    WOR *vs.* time plot, 142
Well MR-321
    Cartesian rate *vs.* time plot, 149, 150
    cumulative production, 149, 150
    derivative plot, 144, 146
    oil and water producing rates, 149, 151
    performance plots, 146
    producing history, 144, 145l
    in rate *vs.* time plot, 149, 150
    total flow through system, 149, 151
west Texas hydraulically fractured oil well, performance history, 27
west Virginia Well A analysis problem
    Cartesian rate *vs.* cumulative production, 42, 43
    flow rate cumulative production plot, 45
    logarithmic rate-logarithmic time plot, 44
    logarithmic rate *vs.* log time, 42, 43
    log flow rate *vs.* time plot, 45
    log rate *vs.* time, 42, 43
    quadratic equation, 42, 44
    quadratic plot, 45
    summary analysis, 44